Springer Theses

Recognizing Outstanding Ph.D. Research

Aims and Scope

The series "Springer Theses" brings together a selection of the very best Ph.D. theses from around the world and across the physical sciences. Nominated and endorsed by two recognized specialists, each published volume has been selected for its scientific excellence and the high impact of its contents for the pertinent field of research. For greater accessibility to non-specialists, the published versions include an extended introduction, as well as a foreword by the student's supervisor explaining the special relevance of the work for the field. As a whole, the series will provide a valuable resource both for newcomers to the research fields described, and for other scientists seeking detailed background information on special questions. Finally, it provides an accredited documentation of the valuable contributions made by today's younger generation of scientists.

Theses are accepted into the series by invited nomination only and must fulfill all of the following criteria

- They must be written in good English.
- The topic should fall within the confines of Chemistry, Physics, Earth Sciences, Engineering and related interdisciplinary fields such as Materials, Nanoscience, Chemical Engineering, Complex Systems and Biophysics.
- The work reported in the thesis must represent a significant scientific advance.
- If the thesis includes previously published material, permission to reproduce this must be gained from the respective copyright holder.
- They must have been examined and passed during the 12 months prior to nomination.
- Each thesis should include a foreword by the supervisor outlining the significance of its content.
- The theses should have a clearly defined structure including an introduction accessible to scientists not expert in that particular field.

More information about this series at http://www.springer.com/series/8790

Sebastian David Stolwijk

Spin-Orbit-Induced Spin Textures of Unoccupied Surface States on Tl/Si(111)

Doctoral Thesis accepted by
the University of Münster, Germany

Springer

Author
Dr. Sebastian David Stolwijk
Physikalisches Institut
University of Münster
Münster
Germany

Supervisor
Prof. Markus Donath
Physikalisches Institut
University of Münster
Münster
Germany

ISSN 2190-5053 ISSN 2190-5061 (electronic)
Springer Theses
ISBN 978-3-319-38703-1 ISBN 978-3-319-18762-4 (eBook)
DOI 10.1007/978-3-319-18762-4

Springer Cham Heidelberg New York Dordrecht London

Softcover reprint of the hardcover 1st edition 2015

Printed on acid-free paper

Springer International Publishing AG Switzerland is part of Springer Science+Business Media
(www.springer.com)

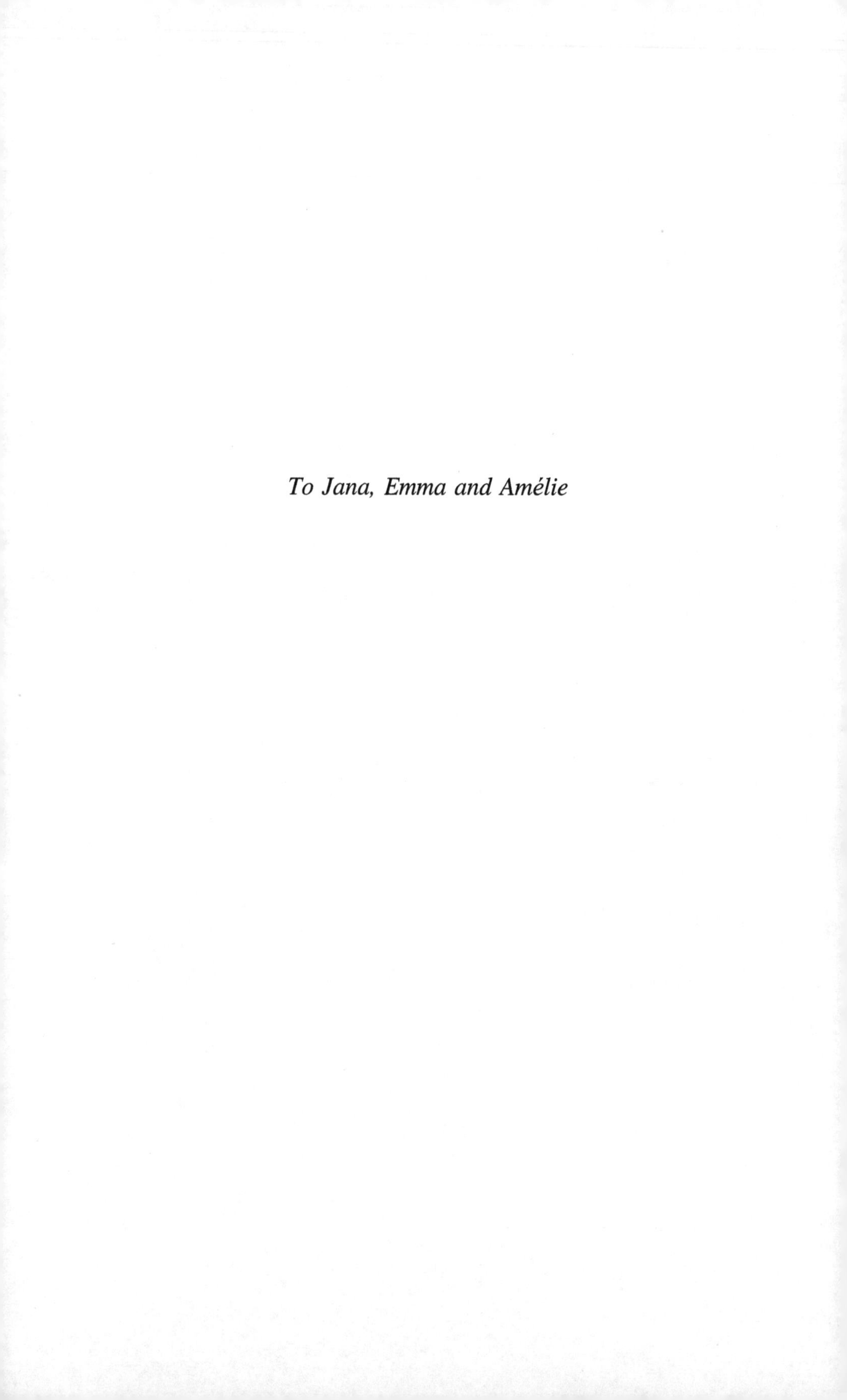

To Jana, Emma and Amélie

You are responsible for your rose.

Antoine de Saint-Exupery, Le Petit Prince
(The Little Prince)

Supervisor's Foreword

The term *spintronics* is used in information technology, where the spin of the electron plays a functional role in electronic devices. Research in this direction aims at using the electron spin as information carrier in addition to the electron charge. Lifting the spin degeneracy of electronic states, which is a necessary condition for reaching this goal, can be achieved by magnetic exchange interaction or by spin-orbit interaction. While in ferromagnetic solids the magnetization direction is a unique quantization axis for the spin, the situation is much more sophisticated in spin-orbit-influenced systems. Here, the preferred spin direction depends not only critically on the wave vector **k** of the electrons but also on the crystal symmetry of the material. A prerequisite for lifting the spin degeneracy is a broken inversion symmetry, which appears, in particular, at the surface. Therefore, the current scientific interest is strongly focused on surface systems comprising heavy elements with strong spin-orbit interaction: Rashba systems and topological insulators. In view of applications in electronic devices, thin layers of heavy metals on top of semiconducting substrates are of special interest, e.g., thallium on silicon, which is the topic of this thesis.

For the Tl/Si(111) system, a number of occupied and unoccupied surface states had been predicted, which, for some **k** values, come close to the Fermi level and, therefore, are interesting for transport properties. The occupied states were intensively studied by spin- and angle-resolved photoemission. Along a specific high-symmetry line, a remarkable rotation of the spin direction as a function of the wave vector was detected. However, the calculations showed that the really interesting surface states close to the Fermi level, which exhibit a giant spin-dependent energy splitting, are unoccupied and, therefore, not accessible to conventional photoemission. To detect spin textures in the unoccupied states for a wide range in reciprocal space, an inverse photoemission apparatus is necessary in combination with a source for spin-polarized electrons, whose spin direction can be tuned. This is the starting point of the thesis at hand. A spin-polarized electron source was modified in such a way that the transversal spin polarization can be tuned freely. This gives experimental access to the classical in-plane Rashba and, in addition, for nonnormal electron incidence of the sample, to the out-of-plane spin-polarization direction. With this source, the unoccupied electron states of one atomic layer of Tl on Si(111) were investigated with respect to their complex spin texture.

The thesis starts with an introduction, which addresses the theme of the work and provides an introduction to the topic of spin-orbit interaction in solids and at surfaces. Chapter 2 is dedicated to the experiment with particular emphasis on the rotatable spin-polarized electron source, the ROSE. It is described and thoroughly characterized by a magnetic system, Ni/W(110), with defined quantization axis and a well-known spin-orbit-influenced Rashba system Au(111). In Chap. 3, the results for the spin-dependent, unoccupied surface electronic structure of Tl/Si(111) are presented. Measurements along high-symmetry directions in **k** space and around high-symmetry points are discussed along with theoretical results:

- Along $\bar{\Gamma}\bar{K}$ and $\bar{\Gamma}\bar{K}'$, the spin-polarization vector rotates from the classical in-plane Rashba polarization direction around $\bar{\Gamma}$ to the direction perpendicular to the surface at the $\bar{K}(\bar{K}')$ points—a direct consequence of the symmetry of the two-dimensional hexagonal system. A giant splitting in energy of about 0.6 eV is observed and attributed to the strong localization of the unoccupied surface state close to the heavy Tl atoms. This leads to completely out-of-plane spin-polarized valleys in the vicinity of the Fermi level. As the valley polarization is oppositely oriented at the $\bar{K}$ and $\bar{K}'$ points, backscattering should be strongly suppressed in this system.
- Along $\bar{\Gamma}\bar{M}$, a surface resonance with giant spin-orbit-induced spin splitting is identified, which exhibits an exclusive Rashba-type spin texture, in compliance with the mirror-plane symmetry of the $\bar{\Gamma}\bar{M}$ direction. For slight deviations from the high-symmetry line, this constraint is lifted and additional spin-polarization components emerge. This highlights the importance of a correct alignment of the experiment when investigating samples with spin textures that are more complex than in simple Rashba systems or in ferromagnets.
- The rotating spin along $\bar{\Gamma}\bar{K}(\bar{\Gamma}\bar{K}')$ and the thin line of a pure Rashba-type spin polarization along $\bar{\Gamma}\bar{M}$ lead to a remarkable spin texture in momentum space around the $\bar{M}$ point. The surface state/resonance encircles the $\bar{M}$ point with a pure in-plane spin polarization at the intersections with the $\bar{\Gamma}\bar{M}$ lines and strong out-of-plane spin polarization at the intersections with the $\bar{M}\bar{K}(\bar{M}\bar{K}')$ lines: a spin texture with a twist around $\bar{M}$.

The interpretation of this very demanding experimental work was made possible by excellent theoretical support (Peter Krüger, Institut für Festkörpertheorie, Westfälische Wilhelms-Universität Münster). The thesis closes with a concise summary, a small supplement, and a comprehensive reference list. The reader is encouraged to discover the beauty of the ROSE.

Münster, Germany
April 2015

Prof. Markus Donath

Abstract

The focus of this work lies in the experimental study of the spin-dependent unoccupied surface electronic structure of Tl/Si(111)-(1×1). Particular emphasis is put on the unoccupied surface states and their spin-orbit-induced spin textures.

The unoccupied surface electronic structure is accessed by means of spin- and angle-resolved inverse photoemission. In order to unveil the spin textures of the unoccupied surface states, a rotatable spin-polarized electron source has been developed, whose key feature is a variable direction of the transversal electron beam polarization. As a result, the spin- and angle-resolved inverse photoemission experiment becomes sensitive to all three spin-polarization directions. This is demonstrated by a thorough characterization of the essential source parameters and two performance tests. Spin- and angle-resolved inverse photoemission measurements of magnetized Ni films on W(110) serve as a reference to demonstrate the variable spin sensitivity. Experiments on Au(111), the prototypical Rashba system, highlight the importance of a correct electron beam and sample alignment, when investigating systems with spin textures that are more complex than systems with a single spin-quantization axis, e.g., ferromagnetic surfaces.

For Tl/Si(111)-(1×1), a spin-orbit-split surface state with unique spin texture is detected throughout the complete surface Brillouin zone. It serves as an ideal sensor to study the intimate interplay between spin-orbit interaction and the surface symmetry. Along $\bar{\Gamma}\bar{K}\,(\bar{\Gamma}\bar{K}')$, the spin-polarization vectors of the surface-state components rotate from the classical in-plane Rashba spin-polarization direction around $\bar{\Gamma}$ to the direction perpendicular to the surface at the $\bar{K}\,(\bar{K}')$ points—a direct consequence of the symmetry of the two-dimensional hexagonal system. Remarkably, at the $\bar{K}\,(\bar{K}')$ points, a giant splitting in energy of about 0.6 eV is observed and attributed to the strong localization of the unoccupied surface state close to the heavy Tl atoms. This leads to completely out-of-plane spin-polarized valleys in the vicinity of the Fermi level. Intriguingly, adsorption of additional Tl renders the valleys metallic. As the valley polarization is oppositely oriented at the $\bar{K}$ and $\bar{K}'$ points, this results in a peculiar Fermi surface, where backscattering is strongly suppressed—a property which may be indispensable for spintronic applications relying on the electron spin as information carrier. The symmetry of the system also determines the spin textures along the other high-symmetry directions. Along $\bar{\Gamma}\bar{\text{M}}$, the surface state becomes a surface resonance, which exhibits a giant spin splitting around $\bar{\text{M}}$. As $\bar{\Gamma}\bar{\text{M}}$ lies in a mirror plane of the system, the surface resonance is purely in-plane spin polarized. Notably, the latter restriction is already lifted for slight deviations from the high-symmetry line. Along $\bar{\text{K}}\bar{\text{M}}$ ($\bar{\text{K}}'\bar{\text{M}}$), the surface state shows a giant spin splitting around $\bar{\text{M}}$ as well, but is, in contrast to the $\bar{\Gamma}\bar{\text{M}}$ direction, predominately out-of-plane spin polarized. In total, this gives rise to a surface state around $\bar{\text{M}}$ with peculiar spin texture, which can be viewed as spin chirality in momentum space.

Acknowledgments

I can honestly state that I enjoyed my time and work as a Ph.D. student. Primarily, this is due to all the great people, who surrounded and supported me in some way or another in carrying out the research and in accomplishing this thesis. I am using this opportunity to express my gratitude.

First of all, I would like to express my special appreciation and thanks to my advisor Prof. Dr. Markus Donath, who gave me the opportunity to work at his chair. You have been a great mentor for me. I would like to thank you for encouraging my research and for allowing me to grow as a research scientist. Your advice on both research as well as on my career has been priceless.

I truly thank Prof. Dr. Kazuyuki Sakamoto for initiating our project on Tl/Si (111). Thank you Kazu for our fruitful collaboration. Without you, this thesis would have been a whole lot different.

Furthermore, I would like to thank Dr. Anke B. Schmidt and Prof. Dr. Peter Krüger for their valuable and essential contributions to the research on Tl/Si(111) and our publications. Thank you for numerous great discussions on physics and the excellent scientific support.

At the heart of this thesis lies the experiment. I thank all, who helped to keep the experiment running. Especially, I would like to thank the technical staff members of the AG Donath Werner Mai and Hubert Wensing for their constant support in any electronic and technical aspect of the laboratory. Without your expertise and help, this work would not have been possible. Furthermore, I thank the members of the workshop of the Physikalisches Institut of the University of Münster for their help in constructing and developing the rotatable spin-polarized electron source, which is a pillar of this work. In this regard, I address special thanks to Werner David and Walter Spiekermann.

The excellent working atmosphere in the AG Donath has been one of the main reasons for my enjoyable Ph.D. time. Therefore, I would like to express my thanks to all the members and former members of the AG Donath research group. A big thanks goes to my colleagues in the lab: Henry Wortelen, Michael Budke, Tobias Allmers, Karen Zumbrägel, Philipp Eickholt, and Bernd Engelkamp.

It is gratefully acknowledged that the University of Münster provided a fruitful research environment.

It goes without saying that this work would have hardly been possible without the support of my family. I warmly thank and appreciate my parents, my mother and father-in-law, my sisters, and my brother and sister-in-law for their support in all aspects of my life. Above all, I thank my beloved wife Jana, who cherished with me every great moment and supported me whenever I needed it. Thank you for being you, accepting me in all ways and making me happy.

Contents

Acronyms

2D	Two-dimensional
3D	Three-dimensional
AES	Auger electron spectroscopy
ARPES	Angle-resolved photoelectron spectroscopy
BIA	Bulk inversion asymmetry
DAS	Dimer adatom stacking fault
DPRF	Differentially pumped rotary feedthroughs
FWHM	Full width at half maximum
hcp	Hexagonal closed-packed
IPE	Inverse photoemission
LEED	Low-energy electron diffraction
ML	Monolayer
MOKE	Magneto-optic Kerr effect
PE	Photoemission
ROSE	Rotatable spin-polarized electron source
SARPES	Spin- and angle-resolved photoelectron spectroscopy
SIA	Structure inversion asymmetry
SOI	Spin-orbit interaction
SPLEED	Spin-polarized low-energy electron diffraction
SPV	Surface photovoltage
SR-IPE	Spin- and angle-resolved inverse photoemisison
STM	Scanning tunneling microscopy
TCS	Target current spectroscopy
UHV	Ultra-high vacuum

Chapter 1
Introduction

1.1 Scope of the Thesis

The notion in the 1930s by Wolfgang Pauli (1900–1958) that "surfaces were invented by the devil" has perished along with the development of ultra-high vacuum (UHV) conditions and sophisticated tools of surface science. Nowadays, the surface of a solid is regarded as a playground for physics with an immense scope of potential subject areas (Plummer et al. 2002). In surface science, considerable attention is paid to surface states—electronic states, which *live* at the surface of a crystalline solid. Surface states reflect the (quasi) two-dimensional (2D) nature of the surface and are ideal sensors to study physics with reduced dimensionality and broken symmetry. In this light, the thesis at hand will address the influence of spin-orbit interaction (SOI) on surface states.

Since the beginnings of quantum mechanics it has been known that spin-orbit interaction (SOI), i.e., the coupling of the electron spin to its orbital momentum, can affect the splitting of energy levels in atoms, molecules and solids. At the crystalline surface, the translational symmetry of the crystal is broken and SOI can lead to spin-polarized surface states even in absence of magnetic fields. In topological insulators, SOI leads to a band inversion, which in turn gives rise to spin-polarized surface states closing the fundamental band gap of the otherwise insulating material. The immense interest in topological insulators is fueled by the unique properties of the topological surface states, which are often believed to revolutionize the use of the electron spin as information carrier in electronic devices, i.e., spintronics. On the other hand, surfaces of metal single crystals and metal adlayer systems may feature surface states, which would also emerge without the presence of SOI. Here, SOI gives rise to a spin splitting of the surface states—a phenomenon, which is often referred to as the Rashba-Bychkov effect (Bychkov and Rashba 1984). Like topological states, Rashba-type spin-orbit-split states may open the way for promising spintronic applications such as the spin field-effect transistor (Datta and Das 1990), the spin Hall-effect transistor (Wunderlich et al. 2010) and others (Schliemann et al. 2003; Fabian et al. 2007). The underlying Rashba-Bychkov model (Bychkov and Rashba 1984) is able

S.D. Stolwijk, *Spin-Orbit-Induced Spin Textures of Unoccupied Surface States on Tl/Si(111)*, Springer Theses, DOI 10.1007/978-3-319-18762-4_1

to describe spin-orbit-induced spin splittings of nearly-free-electron-like dispersing states, where the spin-polarization direction is always in-plane and perpendicular to the momentum parallel to the surface. However, the Rashba-Bychkov model fails to capture the influence of the surface symmetry, which ultimately determines the spin textures of surface states. The spin textures may become far more complex than described by the Rashba-Bychkov model.

In this context, the presented work focuses on the experimental study of Tl/Si(111)-(1×1), a thin heavy-metal film on a semiconducting substrate, to address two inalienable aspects: (i) The search for a metallic surface state with large spin splitting in the fundamental band gap of a semiconductor, where the spin transport is solely determined by the surface state and (ii) the spin-orbit- and symmetry-induced spin textures of the surface states. Standardized tools of surface science such as low-energy electron diffraction (LEED), Auger electron spectroscopy (AES), scanning tunneling microscopy (STM) and angle-resolved photoelectron spectroscopy (ARPES) are used to investigate the surface symmetry, the cleanliness, the morphology and the occupied surface electronic structure of the Tl/Si(111)-(1×1) samples, respectively. The emphasis is on the unoccupied surface states, which are investigated with the central technique employed in this work: spin- and angle-resolved inverse photoemission (SR-IPE). To shed light onto the spin textures of the surface states, the implementation of three-dimensional (3D) spin analysis in the SR-IPE experiment, i.e., the possibility to measure all three spin-polarization components, is a mandatory prerequisite and one of the main objectives of this thesis.

1.2 Physics Background of Spin-Orbit Interaction

Spin-orbit interaction (SOI) is a relativistic effect, which couples the electron's spin degree of freedom to its momentum and gradients of potentials. A detailed discussion of SOI can be found, e.g., in Kessler (1985) and Winkler (2003). The goal of Sect. 1.2.1 is to convey the consequences of SOI for crystalline solids. With the object of investigation, Tl/Si(111)-(1×1), in mind, the priority of Sect. 1.2.2 is set on the crystalline surface. Here, electronic states evolve, which result from the translational symmetry breaking along the surface normal: surface states. It will be shown that with the broken 3D inversion symmetry at the surface, surface states may become spin split. Notably, a spin splitting only occurs if two conditions are fulfilled. (i) A potential gradient mediates the spin-orbit coupling. Typically, the main contribution to the potential gradient is provided by the potential of the atomic core in the vicinity of the surface-state wavefunction. The heavier the element, the stronger is the spin-orbit coupling. (ii) The charge distribution of the surface state is asymmetric along one or more directions. An asymmetry along the surface normal leads to a spin splitting, which can be described by the Rashba-Bychkov model. The stronger the asymmetry, the larger is the spin splitting.

1.2.1 Spin-Orbit Interaction in Crystalline Solids

Briefly, SOI can be understood as a result of the magnetic interaction of the spin magnetic moment $\boldsymbol{\mu}$ with a magnetic field $\mathbf{B}$ in the rest frame of the moving electron, which arises in the presence of an electric field. In the nonrelativistic limit of the Dirac equation, this is expressed by a relativistic correction in terms of a Zeeman-type contribution $H_{SO} = -\boldsymbol{\mu}\mathbf{B}$ to the Hamiltonian of the system. SOI is probably best understood for atoms, where SOI gives rise to the fine-structure splitting of atomic levels. Likewise, SOI is known to have important consequences for the one-electron energy levels in crystalline solids (Dresselhaus 1955; Dresselhaus et al. 2008). In crystalline solids, the one-electron Hamiltonian including SOI is given by

$$H = \frac{\hbar^2\mathbf{k}^2}{2m^*} + V(\mathbf{r}) + \frac{\hbar^2}{4m^{*2}c^2}(\nabla V \times \mathbf{k}) \cdot \boldsymbol{\sigma}, \tag{1.1}$$

where $\hbar$ is the reduced Planck constant, m^* is the effective mass of the electron, c is the velocity of light, $V(\mathbf{r})$ is the lattice potential and $\mathbf{k} = \mathbf{p}/\hbar$ is the wave vector. $\boldsymbol{\sigma}$ is the vector of Pauli matrices and is related to the electron spin $\mathbf{S}$ via $\boldsymbol{\sigma} = \frac{2}{\hbar}\mathbf{S}$. Solving the Schrödinger equation, the electronic states of the crystalline solid can be described by Bloch states $\Psi_{n,\mathbf{k},\chi}(\mathbf{r})$ with the eigenvalues $E_{n,\chi}(\mathbf{k})$, the energy bands. Here, n denotes the quantum number and χ the quantum number of the two spin degrees of freedom, which are denoted as $+$ and $-$, respectively.

Prominent examples to discuss spin-orbit-induced splittings in crystalline solids are crystals with diamond and zinc blende structures, such as Ge and GaAs, respectively. These lattices are equivalent but for the fact that the two-atom basis of the zinc blende lattice consists of two different atoms, i.e., the zinc blende structure is not inversion symmetric. For both structures, SOI leads to a splitting of the topmost valence band (Malone and Cohen 2013). This is, e.g., utilized in GaAs photocathodes to produce spin-polarized electrons via excitation with circularly polarized light (Pierce and Meier 1976). Intriguingly, in diamond structures each band is still at least twofold (spin-) degenerate, whereas in zinc blende structures a k-dependent spin splitting occurs (Dresselhaus 1955; Cardona et al. 1988). How can this difference be understood? In general, the degeneracies or nondegeneracies of the electronic states are governed by (i) the symmetry of the crystal and (ii) the presence or absence of time-reversal symmetry.

(i) The symmetry of the crystal is reflected in point group symmetry operations $\hat{\mathbf{O}}$ of the crystal that leave at least one point of the lattice unmoved and the crystal lattice unchanged $V(\hat{\mathbf{O}}\mathbf{r}) = V(\mathbf{r})$, e.g., rotations, reflections and inversions. An electronic state $\Psi_{n,\mathbf{k},+}(\mathbf{r})$ has then to be invariant under the transition $\mathbf{r} \rightarrow \hat{\mathbf{O}}\mathbf{r}$. Therefore, in inversion-symmetric crystals, where $V(\mathbf{r}) = V(-\mathbf{r})$, the two states $\Psi_{n,\mathbf{k},+}(\mathbf{r})$ and $\Psi_{n,-\mathbf{k},+}(\mathbf{r})$ are degenerate:

$$E_{n,+}(\mathbf{k}) = E_{n,+}(-\mathbf{k}). \tag{1.2}$$

(ii) Time-reversal symmetry denotes the invariance under the transition $t \rightarrow -t$. In other words, time-reversal symmetry preserves the degeneracy under the transition $\mathbf{k} \rightarrow -\mathbf{k}$ with accompanying spin flip: $\Psi_{n,\mathbf{k},+}(\mathbf{r}, t) = \Psi_{n,-\mathbf{k},-}(\mathbf{r}, t)$. The corresponding eigenvalues have to fulfill

$$E_{n,+}(\mathbf{k}) = E_{n,-}(-\mathbf{k}). \quad (1.3)$$

Ultimately, in the case of an inversion-symmetric solid and in the presence of time-reversal symmetry, the Bloch states for every wave vector $\mathbf{k}$ are at least twofold (spin-) degenerate, i.e., $E_{n,+}(\mathbf{k}) = E_{n,-}(\mathbf{k})$. However, if the electrons move in an inversion-asymmetric potential, the spin degeneracy may be lifted: $E_{n,+}(\mathbf{k}) \neq E_{n,-}(\mathbf{k})$. The symmetry of the crystal further determines the arising spin quantization axes. For instance, with a mirror plane in the k_x-k_z plane of the crystal, invariance under the transition $k_y \rightarrow -k_y$ has to be satisfied. Therefore, in the k_x-k_z plane, only the spin degeneracy along k_y is lifted. This restriction does not apply right next to the mirror plane. Altogether, the interplay between SOI and the symmetry of the crystal may give rise to complex spin textures in reciprocal space.

1.2.2 *Spin-Orbit Interaction at Surfaces*

At the crystalline surface, the 3D inversion symmetry of the crystal is broken in any case. As a result, the quasi-2D electronic states of the surface, the surface states, may become spin split. Often, SOI-induced spin splittings of quasi-2D and 2D electron systems are discussed for semiconductor quantum wells in terms of two major contributions (Awschalom et al. 2002; Winkler 2003; Dyakonov 2008): (i) The bulk inversion asymmetry (BIA) relates to the symmetry of the underlying crystal and is also denoted as the Dresselhaus contribution H_D. (ii) The structure inversion asymmetry (SIA), also referred to as Rashba contribution H_R, results from an asymmetry of the confinement potential.

In the simplest approach, a 2D electron system can be described by means of a two-dimensional electron gas (2DEG), i.e., electrons, which are free to move in two dimensions (x and y direction) but are strictly confined in the third (z direction). The electronic bands can then be described by a nearly-free electron dispersion $E(\mathbf{k}_{\parallel})$, where $\mathbf{k}_{\parallel} = (k_x, k_y)$ is the wave vector in the x-y plane. In this case, the symmetry of the crystal structure is irrelevant and BIA-induced splittings vanish. However, a potential gradient $\nabla V = \frac{\partial V}{\partial z}\mathbf{e_z}$ along the confining direction z may lead to a SIA-induced spin splitting. This is often referred to as the Rashba-Bychkov or Rashba effect (Bychkov and Rashba 1984). The Rashba spin splitting of $E(\mathbf{k}_{\parallel})$ is illustrated in Fig. 1.1 and is given by

$$E_{\pm}(\mathbf{k}_{\parallel}) = E_0 + \frac{\hbar^2\mathbf{k}_{\parallel}^2}{2m^*} \pm \alpha_R|\mathbf{k}_{\parallel}|, \quad (1.4)$$

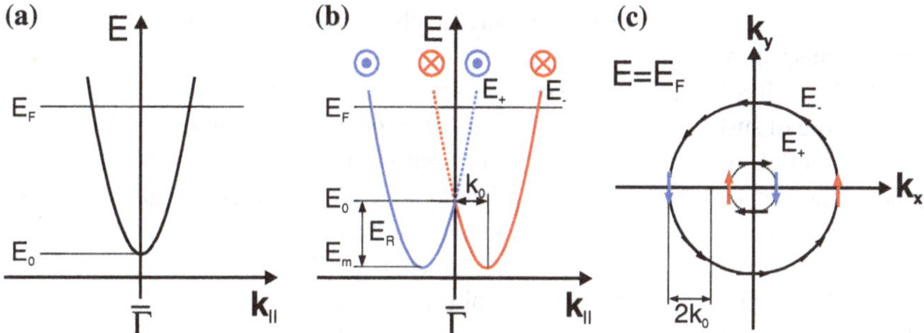

Fig. 1.1 (a) Illustration of a spin-degenerate surface state with free-electron-like dispersion. (b) Spin-orbit interaction lifts the spin degeneracy and two spin-split surface-state components E_+ (*dotted lines*) and E_- (*solid lines*) are formed. The spin-polarization direction (*blue* and *red colored symbols* and *lines*) is always perpendicular to $\mathbf{k}_\parallel$ and lies in the surface plane. The Rashba energy $E_R = |E_0 - E_m|$ is defined as energy offset between the band crossing E_0 and the band extremum E_m. The spin splitting can be interpreted as a linear splitting in $k_\parallel = |\mathbf{k}_\parallel|$ of $2k_0$. k_0 denotes the momentum shift of the band extremum away from the crossing point. (c) Constant-energy contour of the spin-split surface state at $E = E_\mathrm{F}$ (Fermi surface). In the k_x-k_y plane, E_+ and E_- appear as concentric circularly shaped states with antipodal spin-polarization direction

where E_0 denotes the crossing point of the two spin states and E_R the energy difference between the band extremum and the crossing point. The Rashba parameter α_R is a measure of the spin splitting and can be expressed as $\alpha_R = \frac{\hbar^2 k_0}{2m^*}$, where k_0 is the momentum shift of the band extremum away from the crossing point. The quantization axis of the spin polarization lies in the surface plane and is perpendicular to $\mathbf{k}_\parallel$. In the k_x-k_y plane, the state components $E_\pm$ possess a circular shape with antipodal spin polarization (see Fig. 1.1c). The expectation value of the spin polarization is unity and the polarization direction is determined by the sign of α_R and the effective mass m^* (see, e.g., Bentmann et al. 2011).

This rather simple model for strictly 2D states is transferable to quasi-2D surface states, which can be described by $\Psi_S(\mathbf{k}_\parallel) = \exp(i(k_x x + k_y y))\phi(z)$ (Nagano et al. 2009).[1] In this case, the magnitude of the spin splitting E_R relates to

$$\alpha_R \propto \int d\mathbf{r} \frac{\partial V}{\partial z} |\Psi_S(\mathbf{k}_\parallel)|^2. \tag{1.5}$$

Importantly, spin splitting only occurs, if the surface-state wavefunction squared along the z direction $|\phi(z)|^2$, i.e., the charge distribution of the surface state along z, is asymmetric (Nagano et al. 2009). The magnitude of the splitting is determined by the asymmetry of the charge distribution and the Coulomb gradient ($\frac{\partial V}{\partial z}$), which is dominant in the vicinity of the atomic cores (Bihlmayer et al. 2007; Nagano et al. 2009; Bentmann et al. 2011). Therefore, a surface state with localization close to the

[1]For a strictly 2D state one has $\phi(z) = \delta(z - z_0)$, where z_0 is the localization of the surface state along the z direction.

core of a heavy element and with asymmetric charge distribution is likely to possess a large spin splitting.

Actually, it has been demonstrated for several metal surfaces that SOI leads to a spin splitting of surface states, with similar k dependence as predicted by the Rashba-Bychkov model. In particular, good agreement with the Rashba-Bychkov model is found, if the surface state can be described by a nearly-free electron gas. This is the case for, e.g., the surface states of Au and Cu in the vicinity of the $\bar{\Gamma}$ point (LaShell et al. 1996; Hochstrasser et al. 2002; Hoesch et al. 2004; Tamai et al. 2013; Wissing et al. 2013). Notably, the observed spin splittings are of the order 100 meV (Ast et al. 2007; Gierz et al. 2009; Tamai et al. 2013; Stolwijk et al. 2013), orders of magnitude larger than those found in semiconductor heterostructures (0.1–10 meV) (Winkler 2003; Henk et al. 2003).

On the other hand, the Rashba-Bychkov model often fails to describe essential properties of spin-orbit-split surface states, such as their $E(\mathbf{k}_{\parallel})$ dispersions and their spin textures. Indeed, many surfaces exhibit surface states with large anisotropic spin-orbit-induced spin splittings and complex spin textures, which considerably deviate from the predictions of the Rashba-Bychkov model (Ast et al. 2007; Meier et al. 2008; Sakamoto et al. 2009; Mirhosseini et al. 2009; Takayama et al. 2011; Höpfner et al. 2012; Miyamoto et al. 2012a, b). Primarily, this is due to the negligence of the symmetry of the surface, which ultimately determines the dispersion and degeneracies and nondegeneracies of eigenvalues of the given system. This can also be discussed in terms of BIA-related contributions (Höpfner et al. 2012). Inversion-asymmetric surfaces may lead to nonisotropic electric field gradients and in-plane asymmetries of the surface-state wavefunction, which in turn relate to out-of-plane spin polarizations. Therefore, to grasp all aspects of SOI-induced phenomena at surfaces, one is in the need of sophisticated band structure calculations including spin-orbit interaction and, first and foremost, measurements on the surface electronic structure with sensitivity to all spin-polarization directions.

In this light, the unoccupied surface state of Tl/Si(111)-(1×1) can be viewed as a prototypical system to study the interplay between the surface symmetry and SOI. It will be demonstrated that the C_{3v} symmetry of the Tl/Si(111)-(1×1) surface in combination with the strong spin-orbit coupling of the Tl adlayer leads to a complex spin texture in momentum space beyond the simple spin pattern predicted by the Rashba-Bychkov model.

1.3 Outline of the Thesis

The thesis at hand is organized as follows: Chap. 2 covers the experimental aspects. A brief discussion of the principle of SR-IPE is given in Sect. 2.1 and an introduction to the general experimental setup is presented in Sect. 2.2. A main task of this work has been the implementation of 3D spin analysis in SR-IPE with the help of a rotatable spin-polarized electron source—the ROSE. The ROSE will be presented and characterized in Sect. 2.3. The results on Tl/Si(111)-(1×1) are presented in Chap. 3.

After a brief motivation in Sect. 3.1, an introduction to the sample system Tl/Si(111)-(1×1) is given in Sect. 3.2. The preparation of Tl/Si(111)-(1×1) is described and the Tl/Si(111)-(1×1) samples are characterized in terms of surface quality and symmetry aspects. Hereafter, the results on the unoccupied surface states and their spin textures are presented in Sects. 3.3–3.7. It will be demonstrated that the C_{3v} symmetry of the Tl/Si(111)-(1×1) surface in combination with the strong spin-orbit coupling of the Tl adlayer leads to a complex spin texture in momentum space beyond the simple spin pattern predicted by the Rashba-Bychkov model. Finally, a summary of the key results and an outlook on further SR-IPE investigations are given in Chap. 4.

References

Ast, C.R., Henk, J., Ernst, A., Moreschini, L., Falub, M.C., Pacilé, D., Bruno, P., Kern, K., Grioni, M.: Giant spin splitting through surface alloying. Phys. Rev. Lett. **98**, 186807 (2007)

Awschalom, D.D., Loss, D., Samarth, N. (eds.): Semiconductor Spintronics and Quantum Computation. Springer, New York (2002)

Bentmann, H., Kuzumaki, T., Bihlmayer, G., Blügel, S., Chulkov, E.V., Reinert, F., Sakamoto, K.: Spin orientation and sign of the Rashba splitting in Bi/Cu(111). Phys. Rev. B **84**, 115426 (2011)

Bihlmayer, G., Blügel, S., Chulkov, E.V.: Enhanced Rashba spin-orbit splitting in Bi/Ag(111) and Pb/Ag(111) surface alloys from first principles. Phys. Rev. B **75**, 195414 (2007)

Bychkov, Y.A., Rashba, E.I.: Properties of a 2D electron gas with lifted spectral degeneracy. JETP Lett. **39**, 78 (1984)

Cardona, M., Christensen, N.E., Fasol, G.: Relativistic band structure and spin-orbit splitting of zinc-blende-type semiconductors. Phys. Rev. B **38**, 1806 (1988)

Datta, S., Das, B.: Electronic analog of the electro-optic modulator. Appl. Phys. Lett. **56**, 665 (1990)

Dresselhaus, G.: Spin-orbit coupling effects in zinc blende structures. Phys. Rev. **100**, 580 (1955)

Dresselhaus, M.S., Dresselhaus, G., Jorio, A.: Group Theory: Application to The Physics of Condensed Matter. Springer, Berlin (2008)

Dyakonov, M.I. (ed.): Spin Physics in Semiconductors. Springer, Berlin (2008)

Fabian, J., Matos-Abiague, A., Ertler, C., Stano, P., Žutić, I.: Semicond. spintronics. Acta Phys. Slovaca **57**, 565 (2007)

Gierz, I., Suzuki, T., Frantzeskakis, E., Pons, S., Ostanin, S., Ernst, A., Henk, J., Grioni, M., Kern, K., Ast, C.R.: Silicon surface with giant spin splitting. Phys. Rev. Lett. **103**, 046803 (2009)

Henk, J., Ernst, A., Bruno, P.: Spin polarization of the L-gap surface states on Au(111). Phys. Rev. B **68**, 165416 (2003)

Hochstrasser, M., Tobin, J.G., Rotenberg, E., Kevan, S.D.: Spin-resolved photoemission of surface states of W(110)-(1×1)H. Phys. Rev. Lett. **89**, 216802 (2002)

Hoesch, M., Muntwiler, M., Petrov, V.N., Hengsberger, M., Patthey, L., Shi, M., Falub, M., Greber, T., Osterwalder, J.: Spin structure of the Shockley surface state on Au(111). Phys. Rev. B **69**, 241401 (2004)

Höpfner, P., Schäfer, J., Fleszar, A., Dil, J.H., Slomski, B., Meier, F., Loho, C., Blumenstein, C., Patthey, L., Hanke, W., Claessen, R.: Three-dimensional spin rotations at the Fermi surface of a strongly spin-orbit coupled surface system. Phys. Rev. Lett. **108**, 186801 (2012)

Kessler, J.: Polarized electrons. Springer-Verlag, Berlin Heidelberg (1985)

LaShell, S., McDougall, B.A., Jensen, E.: Spin splitting of an Au(111) surface state band observed with angle resolved photoelectron spectroscopy. Phys. Rev. Lett. **77**, 3419 (1996)

Malone, B.D., Cohen, M.L.: Quasiparticle semiconductor band structures including spin-orbit interactions. J. Phys.: Condens. Matter **25**, 105503 (2013)

Meier, F., Dil, H., Lobo-Checa, J., Patthey, L., Osterwalder, J.: Quantitative vectorial spin analysis in angle-resolved photoemission: Bi/Ag(111) and Pb/Ag(111). Phys. Rev. B **77**, 165431 (2008)

Mirhosseini, H., Henk, J., Ernst, A., Ostanin, S., Chiang, C.T., Yu, P., Winkelmann, A., Kirschner, J.: Unconventional spin topology in surface alloys with Rashba-type spin splitting. Phys. Rev. B **79**, 245428 (2009)

Miyamoto, K., Kimura, A., Kuroda, K., Okuda, T., Shimada, K., Namatame, H., Taniguchi, M., Donath, M.: Spin-polarized Dirac-cone-like surface state with *d* character at W(110). Phys. Rev. Lett. **108**, 066808 (2012a)

Miyamoto, K., Kimura, A., Okuda, T., Shimada, K., Iwasawa, H., Hayashi, H., Namatame, H., Taniguchi, M., Donath, M.: Massless or heavy due to two-fold symmetry: surface-state electrons at W(110). Phys. Rev. B **86**, 161411 (2012b)

Nagano, M., Kodama, A., Shishidou, T., Oguchi, T.: A first-principles study on the Rashba effect in surface systems. J. Phys.: Condens. Matter **21**, 064239 (2009)

Pierce, D.T., Meier, F.: Photoemission of spin-polarized electrons from GaAs. Phys. Rev. B **13**, 5484 (1976)

Plummer, E., Ismail, Matzdorf, R., Melechko, A., Pierce, J., Zhang, J.: Surfaces: a playground for physics with broken symmetry in reduced dimensionality. Surf. Sci. **500**, 1 (2002)

Sakamoto, K., Oda, T., Kimura, A., Miyamoto, K., Tsujikawa, M., Imai, A., Ueno, N., Namatame, H., Taniguchi, M., Eriksson, P.E.J., Uhrberg, R.I.G.: Abrupt rotation of the Rashba spin to the direction perpendicular to the surface. Phys. Rev. Lett. **102**, 096805 (2009)

Schliemann, J., Egues, J.C., Loss, D.: Nonballistic spin-field-effect transistor. Phys. Rev. Lett. **90**, 146801 (2003)

Stolwijk, S.D., Schmidt, A.B., Donath, M., Sakamoto, K., Krüger, P.: Rotating spin and giant splitting: unoccupied surface electronic structure of Tl/Si(111). Phys. Rev. Lett. **111**, 176402 (2013)

Takayama, A., Sato, T., Souma, S., Takahashi, T.: Giant out-of-plane spin component and the asymmetry of spin polarization in surface rashba states of bismuth thin film. Phys. Rev. Lett. **106**, 166401 (2011)

Tamai, A., Meevasana, W., King, P.D.C., Nicholson, C.W., de la Torre, A., Rozbicki, E., Baumberger, F.: Spin-orbit splitting of the Shockley surface state on Cu(111). Phys. Rev. B **87**, 075113 (2013)

Winkler, R.: Spin-Orbit Coupling Effects in Two-Dimensional Electron and Hole Systems. Springer, Berlin (2003)

Wissing, S.N.P., Eibl, C., Zumbülte, A., Schmidt, A.B., Braun, J., Minár, J., Ebert, H., Donath, M.: Rashba-type spin splitting at Au(111) beyond the Fermi level: the other part of the story. New J. Phys. **15**, 105001 (2013)

Wunderlich, J., Park, B.G., Irvine, A.C., Zârbo, L.P., Rozkotová, E., Nemec, P., Novák, V., Sinova, J., Jungwirth, T.: Spin hall effect transistor. Science **330**, 1801 (2010)

Chapter 2
Experiment

In the center of this thesis lies the unoccupied surface state of Tl/Si(111)-(1 × 1) and its spin texture. Experimental access to the unoccupied surface state is obtained via spin- and angle-resolved inverse photoemission. To shed light onto the spin texture, which forms complex patterns in momentum space, one of the project's main objectives has been to implement 3D spin analysis in the SR-IPE experiment to get access to all components of the spin polarization.

SR-IPE (Donath 1994, 1999) is the complementary method to spin- and angle-resolved photoelectron spectroscopy (SARPES) (Johnson 1997; Heinzmann and Dil 2012; Okuda and Kimura 2013). In SARPES, spin resolution is obtained in the detection channel with the help of spin-polarization detectors based on spin-orbit interaction, e.g., Mott scattering (Burnett et al. 1994) and spin-polarized low-energy electron diffraction (SPLEED) (Kirschner and Feder 1979; Yu et al. 2007), or exchange interaction, e.g., in very-low-energy electron diffraction detectors (Winkelmann et al. 2008; Okuda et al. 2008; Escher et al. 2011). In SR-IPE, spin resolution is achieved in the excitation channel by the use of spin-polarized electrons. Thus, to implement 3D spin analysis, spin-polarized electrons with variable polarization direction are needed.

In analogy to SARPES, where 3D spin detection is realized by using two spin-polarization detectors (Hoesch et al. 2002; Okuda et al. 2011), one could think of installing an additional electron source, but this requires a complicated setup with two lasers and two photocathodes. An alternative approach for spin manipulation has been realized before. Magnetic fields are used to alter the electron beam polarization direction, e.g., in spin-polarized electron sources for appearance potential spectroscopy (Rangelov et al. 2000) or spin-polarized low-energy electron microscopy experiments (Duden and Bauer 1995). This, however, is disadvantageous, in that magnetic stray fields affect the angular distribution of the electron beam and, as a consequence, limit the momentum resolution. Hence, spin-polarized electron sources

---

Figures and text excerpts of this chapter have been published in Stolwijk et al. (2014). Reprinted with permission from Stolwijk et al. (2014). Copyright 2014, AIP Publishing LLC.

S.D. Stolwijk, *Spin-Orbit-Induced Spin Textures of Unoccupied Surface States on Tl/Si(111)*, Springer Theses, DOI 10.1007/978-3-319-18762-4_2

for inverse photoemission have been restricted to one spin-polarization direction, so far. In this work, a **RO**tatable **S**pin-polarized **E**lectron source (ROSE) has been developed, which enables 3D spin analysis by embedding a variable rotation of the source chamber. It produces a transversally spin-polarized electron beam, whose polarization direction can be chosen freely. This allows investigations of the spin-dependent unoccupied electronic structure with sensitivity to two in-plane spin-polarization directions and, for nonnormal electron incidence, with sensitivity to the out-of-plane spin-polarization component. Importantly, as the pressure in the source chamber during rotation stays below 5×10^{-10} mbar, measurements with different sensitivities can be conducted at one and the same sample preparation.

In the following, an introduction to the basic concept of SR-IPE is given in Sect. 2.1. General aspects of the experimental setup and the most relevant techniques employed in this thesis will be briefly addressed in Sect. 2.2. The ROSE is presented in Sect. 2.3. Constructional details of the ROSE are described in Sect. 2.3.1, whereas the experimental characterization of the ROSE is part of Sect. 2.3.2. Finally, the performance of the SR-IPE experiment with the ROSE is demonstrated on the basis of two examples in Sect. 2.4: SR-IPE of ferromagnetic Ni/W(110) (Sect. 2.4.1) and SR-IPE of the prototypical Rashba-system Au(111) (Sect. 2.4.2).

2.1 Spin- and Angle-Resolved Inverse Photoemission

Inverse photoemission (IPE) can be viewed as Bremsstrahlung spectroscopy, where a solid is bombarded with low-energy electrons (Dose 1977, 1985; Smith 1988). IPE is a form of spontaneous emission and is often regarded as the time-reversed process to photoemission (PE). An electron in an excited state of the system decays into a state with lower energy and emits a quantum of energy of light. In IPE experiments, this is utilized to investigate the unoccupied electronic structure $E(\mathbf{k})$ of solids (Pendry 1980). In SR-IPE experiments, information on the spin-dependent surface electronic structure is obtained by resolving the energy E_{kin}, the angle of electron incidence θ, which is related to the momentum $\mathbf{k}$, and the spin polarization $\mathbf{P}$ of the electrons. SR-IPE becomes surface sensitive by the use of electrons with kinetic energies in the electron-volt range ($E_{\text{kin}} = 7 - 20\,\text{eV}$), because of the shortness ($\approx$10 Å) of the corresponding electron mean free path (Powell 1988). A detailed description of IPE and SR-IPE can be found in Dose (1985), Smith (1988), Himpsel (1990), Donath (1994) and Donath (1999). The most relevant aspects of IPE and SR-IPE for this thesis will be briefly described in the following:

The IPE process is schematically shown in Fig. 2.1a. An impinging electron with defined energy E_{kin} and momentum $\mathbf{k}$ couples to an unoccupied state of the sample, also referred to as initial state $|i\rangle$. Coupling is efficient, if the overlap of the free-electron plane wave and the wave function of the initial state is large. This explicitly depends on the symmetry of the initial state and is strongest for bands with nearly-free electron character (Dose 1985). The electron may then radiatively decay into a lower lying state, the final state $|f\rangle$. An IPE transition can be described within the

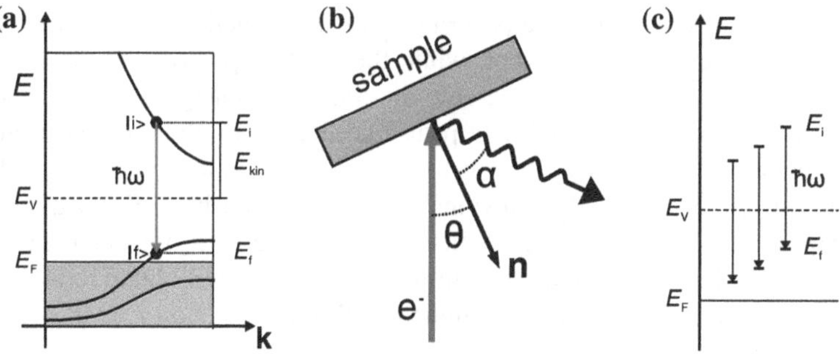

Fig. 2.1 **a** Illustration of the IPE transition in the reduced zone scheme. A free electron couples to an unoccupied state of the crystal, the initial state $|i\rangle$ and radiatively decays into a lower lying unoccupied state, the final state $|f\rangle$. The emitted photon has an energy of $\hbar\omega = E_i - E_f$. **b** Definition of the angle of electron incidence θ and the photon emission angle α with respect to the surface normal **n**. **c** Visualization of the isochromat mode of the IPE experiment. The initial state energy E_i is swept, while the photon detection energy $\hbar\omega$ is kept constant. As a result, the experiment scans the final state energy E_f

dipole approximation. A specific dipole transition relates to a specific dipole axis, which leads to a distinct angular distribution of the photon emission (Eberhardt and Himpsel 1980; Desinger et al. 1984; Donath et al. 1986; Fauster et al. 1989). The transition probability from an initial state $|i\rangle$ to a final state $|f\rangle$ can be expressed via a transition matrix $\langle f|\mathbf{A}\cdot\mathbf{p}|i\rangle$, where $\mathbf{A}$ is the vector potential of the electromagnetic wave and $\mathbf{p}$ is the momentum operator (Eberhardt and Himpsel 1980).

Considering the conditions of conservation of energy, the energy E_i of the initial state is in a fixed relation to the energy of the emitted photon $E_{\text{ph}} = \hbar\omega$ and the final state energy E_f:

$$\hbar\omega = E_i - E_f = E_{\text{kin}} + E_V - E_f, \tag{2.1}$$

where E_i can be expressed by the kinetic energy E_{kin} of the free electron and the vacuum level E_V of the sample ($E_V - E_F$ is also referred to as the work function of the sample, where E_F is the Fermi level). Regarding the conservation of momentum, the momentum parallel to the surface plane $\mathbf{k}_{\|}$ is conserved but for reciprocal lattice vectors of the surface $\mathbf{g}_{\|}$. For $\mathbf{g}_{\|} = 0$, the absolute value of $\mathbf{k}_{\|}$ can be expressed as

$$k_{\|} = \sqrt{\frac{2m_e}{\hbar^2}} \cdot \sqrt{E_{\text{kin}}} \cdot \sin\theta = \sqrt{\frac{2m_e}{\hbar^2}} \cdot \sqrt{\hbar\omega + E_f - E_V} \cdot \sin\theta, \tag{2.2}$$

where m_e denotes the mass of a free electron, $\hbar$ the reduced Planck constant and θ the angle of electron incidence (see Fig. 2.1b). The momentum perpendicular to the surface $\mathbf{k}_{\perp}$ is affected by the vacuum-solid boundary and is not conserved.

Two-dimensional states such as surface states are independent of $\mathbf{k}_{\perp}$ and are completely characterized by the $E(\mathbf{k}_{\|})$ dispersion. This is directly accessible by the IPE experiment and can be obtained as follows. In this work, the IPE experiments are operated in the isochromat mode (see Fig. 2.1c). The kinetic energy of the incident electrons and, thereby, E_i is swept, while the emitted photons are detected with a fixed mean detection energy. Thereby, the final state energy E_f is scanned. An IPE transition may be observed as a peak-like feature in photon yield (intensity) *versus* kinetic energy curves, in analogy to energy distribution curves in photoelectron spectroscopy experiments. By analyzing the energetic position E_f of the surface-state emission for different angles of electron incidence, $E(\mathbf{k}_{\parallel})$ is obtained with the help of Eq. 2.2. Here, three independent methods are used to obtain reliable values for E_f. (i) The positions E_f are determined by fitting a function $I = W * [(\sum_i L_i + B)(1 - f(E))]$ to the SR-IPE spectra, where W is the apparatus function, $f(E)$ the Fermi distribution, L_i the considered Lorenzian-shaped state emissions and B the background intensity (see also Stolwijk et al. 2010). The fit parameters are the energetic positions and the widths of the state emissions. Appropriate backgrounds are chosen taking also into account states at higher/lower energies. Variations of the background are considered to get a measure of the confidence interval. (ii) The minima of the second derivatives of the SR-IPE spectra are analyzed. Ideally, their energetic positions correspond to the energetic positions of the state emissions (see also Appendix A.2). Note that this procedure only leads to reasonable results as long as the background intensity can be treated as a polynomial up to the order E^2. (iii) The positions E_f are determined by an estimation based on the experience of the experimentalist.

In systems where the spin degeneracy of the states is removed, e.g., due to ferromagnetism or SOI (see Sect. 1.2), the respective spin states $\mathbf{S}$ have to be considered. The spin-polarization expectation value $\overline{P} = \langle \mathbf{S} \rangle$ of $|f\rangle$ and $|i\rangle$ is generally less than unity (Henk et al. 1996), whereas $\overline{P}$ of the free electron is unity. Here, $\overline{P}$ is not to be mistaken as the spin-polarization value of the electron beam P, which is defined for the electron ensemble and is, typically, less than 100 %. The spin-polarization direction of the impinging electrons determines the spin sensitivity of the SR-IPE experiment. In this context, an important aspect in SR-IPE measurements deals with the presentation of the gathered data. To account for an incomplete electron beam polarization P, the raw data have to be normalized to 100 % beam polarization. Additionally, one has to consider the angle Φ between the quantization axis of the electron spin polarization and the spin quantization axis of the sample. The raw data are then normalized to 100 % effective beam polarization, see for instance in Donath (1994). If $N_\uparrow$ and $N_\downarrow$ are the raw data of the two spin channels, the normalized data $I_\uparrow$ and $I_\downarrow$ are determined by

$$I_{\uparrow,\downarrow} = \frac{N_\uparrow + N_\downarrow}{2} \cdot \left(1 \pm \frac{N_\uparrow - N_\downarrow}{N_\uparrow + N_\downarrow} \cdot \frac{1}{P \cos \Phi}\right).$$

Normalization is not possible, if Φ is not known. In the case that the measurement is simultaneously sensitive to two spin-polarization directions, additional information is mandatory. Additional information from other experiments or symmetry arguments can help to exclude the existence of one of the possible polarization directions. Note that for SARPES experiments the same considerations apply. There, the raw data is normalized to the effective Sherman function S of the spin-polarization detector.

Conclusively, 3D spin analysis in SR-IPE has to be achieved in the excitation channel by the use of electron beams with variable spin-polarization direction. The following sections are concerned with the experimental realization.

2.2 Experimental Setup

The employed SR-IPE experiment is part of a multi-chamber ultra-high vacuum system, which is schematically shown in Fig. 2.2. Several publications give a detailed description and demonstrate the performance of this setup (Yu et al. 2007; Budke et al. 2007a, b; Budke and Donath 2008, 2009; Allmers and Donath 2009; Stolwijk et al. 2010; Allmers and Donath 2010; Allmers et al. 2011). Basically, it consists of three units (see Fig. 2.2a): the preparation chamber (P), the microscopy chamber (M) and the analysis chamber (A). The chambers are connected via a sample transfer system.

The preparation chamber is equipped with standard tools of surface science for film growth and film characterization. It provides several evaporators and an ion sputter gun. The manipulator accepts Omicron flag-style sample plates and allows sample

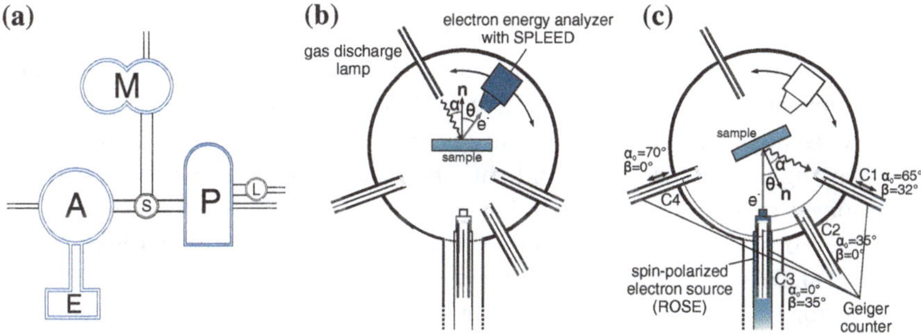

Fig. 2.2 **a** Schematic drawing (not to scale) of the multi-chamber ultra-high vacuum system with the preparation chamber (*P*), the microscopy chamber (*M*), the analysis chamber (*A*) and the electron source chamber (*E*). The chambers are connected via a sample transfer system. Samples can be inserted into the system via a load lock (*L*) and stored in UHV (*S*). **b** (S)ARPES equipment (not to scale). The (S)ARPES experiment consists of a gas discharge lamp and an electron-energy analyzer/SPLEED combination. The polar (azimuthal) angle of photon incidence with respect to the surface normal **n** is $\alpha = 30°$ ($\beta = 35°$). The electron-energy analyzer is mounted on a goniometer to vary the angle of electron emission θ. **c** The SR-IPE setup (not to scale) includes the spin-polarized electron source (shown is only the last part of the electron-transfer optics) and four Geiger counters ($C1$, $C2$, $C3$ and $C4$). The photon detectors are fixed, while the sample is rotated to vary the angle of electron incidence θ. The polar angle of photon detection α of the counters depends on θ. For $\theta = 0°$, the photon detection angles are denoted as α_0. β denotes the azimuthal angle of photon detection. Data shown in this thesis were mainly detected with counter $C1$

cooling to 115 K with liquid nitrogen and indirect heating up to 1200 K. To enable fast heating to high temperatures (flashing) up to 2500 K, the chamber is equipped with an electron bombardment station (Stolwijk 2009). Moreover, the preparation chamber includes sample characterization techniques such as low-energy electron diffraction (LEED) (Kittel 1967; Kopitzki 1989; Henzler and Goepel 1994), medium-energy electron diffraction (MEED),[1] Auger electron spectroscopy (AES) (Auger 1925; Linsmeier 1994) and an in-situ magneto-optical Kerr-effect (MOKE) setup (Kerr 1877; Polisetty et al. 2008; Allmers and Donath 2010).

The microscopy chamber features a scanning tunneling microscope from Omicron NanoTechnology GmbH. A detailed introduction to STM can be found in Binnig and Rohrer (2000) and Chen (2008). In this work, STM is used to characterize the topography of the studied films.

At the heart of the experiment lies the analysis chamber with a unique combined setup for direct and inverse spin- and angle-resolved photoemission (Budke et al. 2007a). In the course of this work, the analysis chamber has been equipped with a 5-axis manipulator from VG Scienta (Cryoax 5 with Omniax 200 mm Z slide module and ±25 mm XY stage) with nonmagnetic Omicron flag-style sample holder (SH2). It enables cooling of the sample to 60 K or to 120 K by the use of liquid helium or liquid nitrogen, respectively. Indirect heating to 1200 K is possible. A key feature of the manipulator is the possibility to rotate the sample about two axes. Rotation about the manipulator axis is denoted as primary rotation θ. Rotation about the surface normal varies the azimuth φ of the sample. The equipment for the (i) SARPES and (ii) SR-IPE experiment is schematically illustrated in Fig. 2.2b, c, respectively.

(i) The SARPES setup consists of a HIS 13 gas discharge lamp from Focus GmbH (Schönhense and Heinzmann 1983) and an electron-energy analyzer (50 mm hemispherical sector analyzer SHA50 from Focus GmbH). The analyzer is mounted on a goniometer to change the electron-emission angle without changing the incidence angle of light. The experiment can be operated in a spin-resolved (SARPES) and a spin-integrated mode (ARPES). Spin resolution is achieved with a SPLEED detector (Focus GmbH) (Kirschner and Feder 1979; Yu et al. 2007) attached to the analyzer. The Sherman function S of the SPLEED detector is about $S = 0.24$ (Budke et al. 2007a). The energy resolution ΔE_{PE} of the ARPES/SARPES experiment depends on the pass energy of the electron analyzer and ranges from 50 to 160 meV full width at half maximum (FWHM). The angular resolution $\Delta\theta_{\text{PE}}$ is approximately $\Delta\theta = 2°$ (FWHM) (Budke et al. 2007a; Allmers 2009).

(ii) The SR-IPE experiment includes four bandpass photon detectors ($C1$-$C4$), i.e., Geiger counters using acetone as counting gas and CaF_2 entrance windows (Budke et al. 2007b; Thiede 2010). This combination creates an optical bandpass, which depends on the temperature T_{CaF_2} of the CaF_2 entrance window. At room temperature, a bandpass of about 320 meV (FWHM) with a mean detection energy of 9.9 eV is obtained. Two detectors ($C2$ and $C4$) possess heatable

[1] Similar to high-energy electron diffraction (Peng et al. 2011).

CaF_2 entrance windows to optimize the total energy resolution of the SR-IPE experiment. A characterization of the bandpass and the energy resolution will be shown in Sect. 2.3.2. Additionally, two detectors ($C1$ and $C4$) are mounted on linear drives to allow a close approach to the sample. Thereby, the solid angle of photon collection is enlarged and the photon yield increased. To measure angular-dependent photon yields, the detectors are positioned at various angles with respect to the surface normal as denoted in Fig. 2.2c. The spin-polarized electron source for the SR-IPE experiment, the ROSE, is located in the source chamber (E), which is connected with the analysis chamber via electron transfer optics. A detailed discussion of the ROSE is presented in the following section.

2.3 The Rotatable Spin-Polarized Electron Source

2.3.1 Constructional Details and Quantization Axes

Briefly, the principle of the ROSE is as follows. A circularly polarized laser beam excites spin-polarized electrons at an activated GaAs photocathode as described in Pierce and Meier (1976) and Pierce et al. (1980). These are extracted by an emission lens and guided through a toroidal electrostatic deflector with a slit aperture at the focal point of the deflector exit. The deflector and slit form the energy-selective element. Additionally, the spin polarization of the electron beam is changed from initially longitudinal to transversal. After passing the slit aperture, the electrons enter the transfer part of the electron optics and are guided through a gate valve into the analysis chamber onto the sample. The gate valve is located in the middle of the transfer optics and can separate the source chamber from the analysis chamber. This is necessary to avoid contamination of the sample during the preparation procedure of the GaAs photoemitter and vice versa.

Up to now, spin-polarized electron sources for inverse-photoemission experiments provided only one fixed polarization direction, with the polarization vector being either perpendicular (Wissing et al. 2013) or parallel (Donath 1994) to the plane of incidence, as shown in Fig. 2.4c, d, respectively. Here, a straightforward way to establish an in-situ variation of the transverse polarization of the electron beam is realized by introducing a rotatable source chamber. For this purpose, two major modifications have been applied to the previous electron source (Budke et al. 2007b).

(i) To implement the gate valve within the electron transfer optics, in previous electron sources a removable screening tube at lens potential was used to bridge the gap between the two parts of the lens element E_2 (see left hand side of Fig. 2.3a and Kolac et al. 1988). In the course of the electron source reconstruction, this mechanical system was replaced by two graphite-coated circular condenser plates at the ends of the lens element E_2, as shown in Fig. 2.3a (right-hand side). This has two advantages: First, the mounting of a removable screening tube and its supporting guide rail in combination with rotatable optic parts is not a trivial task. Second, the

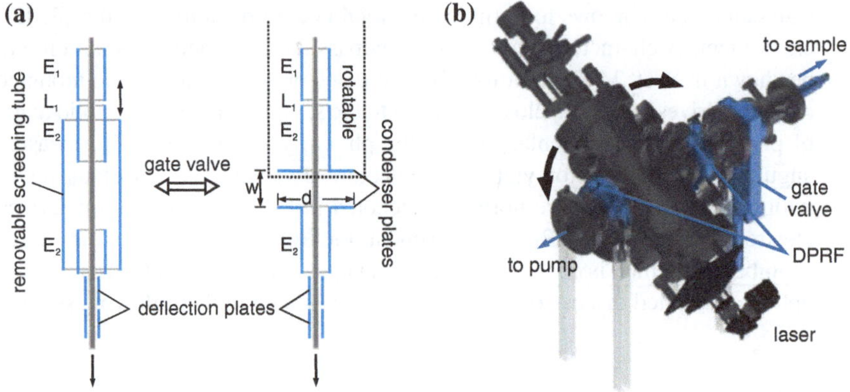

Fig. 2.3 **a** Schematic drawing of the transfer optics of the ROSE in the vicinity of the gate valve. Former (*left-hand side*) and new design (*right-hand side*) with screening tube and condenser plates, respectively. The upper part of the electron optics (indicated by the *dotted box*) is rotated along with the source chamber. **b** Technical drawing of the ROSE. The differentially pumped rotary feedthroughs (DPRF), the gate valve and the electron optics are colored in *blue*. The source chamber can be rotated along the transfer-optics axis as illustrated by the *black arrows*

new design abstains from moving mechanical parts, which may become a source of failure. To ensure a homogeneous electrical field in the region of the electron beam, the diameter of the condenser plates is maximized ($d = 60$ mm) and the distance between the condenser plates is as small as possible ($w = 27$ mm), while still leaving enough room for the gate-valve plate. The beam properties were left unchanged by this modification. At the same time, the gap in the transfer optics constitutes a constructional separation and, therefore, a rotation of the two parts relative to each other is unproblematic. In the setup presented in the scope of this thesis, the optics part attached to the source chamber is rotated with the source chamber, while the other part remains fixed.

(ii) The source chamber is attached between two differentially pumped rotary feedthroughs (DPRF) as shown in Fig. 2.3b, which carry the load of the source chamber. It should be noted that this setup demands for a highly accurate alignment of the source-chamber flanges and rotary drives, as irregular axial stress would likely break the sealing of the differentially pumped rotary drives. Therefore, prior to mounting, the source chamber flanges have been mechanically aligned to be parallel with an accuracy of below 0.1°. In addition, two compensation elements in the form of bellows are embedded to equalize small tilts. As a result, one is able to perform a continuous rotation of the source chamber around the transfer optics axis. Since the part of the transfer optics up to the gate valve is attached to the rotary drive, it is rotated as well. This motivated the modification described in (i). The rotary drives are attached to a turbo-molecular pump/rotary pump combination to ensure UHV conditions in the source chamber. The base pressure of the electron source is less than 1×10^{-10} mbar and stays below 5×10^{-10} mbar during rotation. Hence, the GaAs

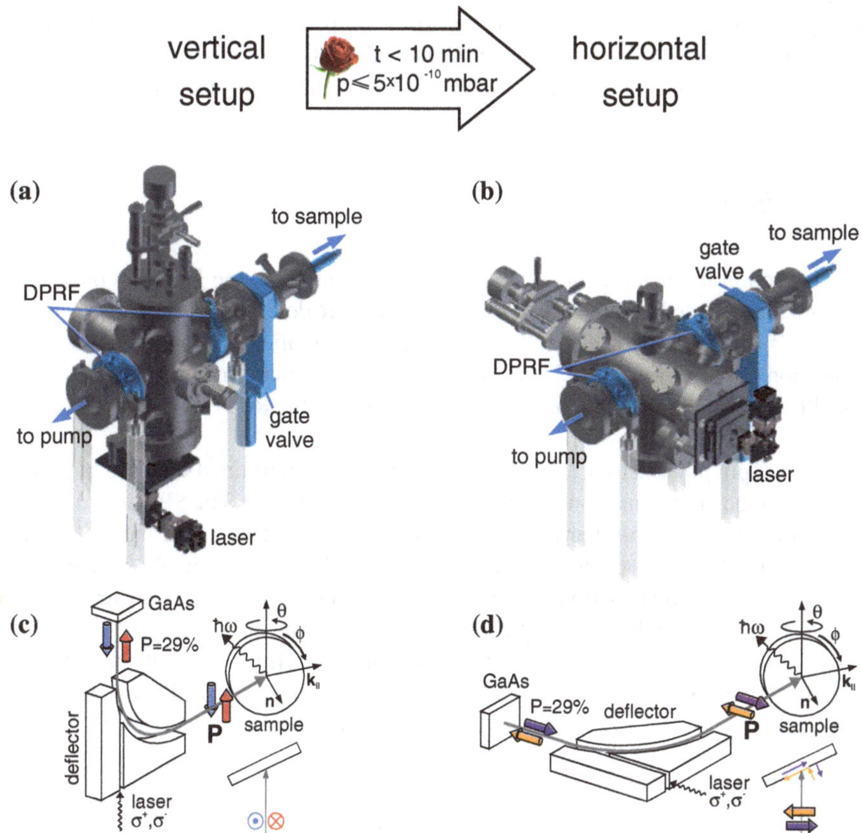

Fig. 2.4 **a**, **b** Technical drawing and **c**, **d** schematic view of the ROSE in (**a**, **c**) vertical and (**b**, **d**) horizontal setup. The schematic representation of the **c** vertical and **d** horizontal setup of the ROSE illustrates the spin-polarization direction of the electron beam, which is perpendicular (vertical setup) or parallel (horizontal setup) to the plane of incidence. Sensitivity to the in-plane spin-polarization direction, i.e., parallel to the surface and perpendicular (vertical setup) or parallel (horizontal setup) to the plane of incidence is obtained. Additionally, in the horizontal setup for $\theta \neq 0°$, out-of-plane sensitivity is realized

photocathode and, for the case that the gate valve is open, the sample condition remain unaffected.

By selecting the vertical and horizontal orientation of the electron source as shown in Fig. 2.4a, b, respectively, it is possible to analyze two orthogonal in-plane spin-polarization directions. This becomes clear in the schematic view of the vertical and horizontal setup (see Fig. 2.4c, d). In the vertical (horizontal) setup, the SR-IPE experiment is sensitive to the in-plane spin-polarization direction perpendicular (parallel) to the plane of incidence, which is denoted as in-plane$_{\perp}$ (in-plane$_{\parallel}$) direction. In the case of horizontal orientation and nonnormal electron incidence, i.e., $\theta \neq 0°$, the experiment is simultaneously sensitive to the in-plane$_{\parallel}$ and the out-of-plane spin-polarization component (see polarization components in Fig. 2.4d). Switching between horizontal and vertical position takes about five to ten minutes time. After rotation, only minor corrections have to be made to the electron beam alignment.

2.3.2 Characterization

In this section the inverse-photoemission experiment using the ROSE is put to the test and the essential parameters of the experiment are determined: (i) the spin polarization P, (ii) the width of the electron-energy distribution $\Delta E_{\rm el}$, (iii) the angular resolution $\Delta\theta$, (iv) the spot size $d_{\rm el}$ and (v) the total energy resolution ΔE of the SR-IPE experiment. It is emphasized that a rotation of the ROSE does not vary these values.

(i) The spin polarization P of the electron beam is determined with the SPLEED detector of the photoemission setup as sketched in Fig. 2.5a, b. The SPLEED detector is sensitive to the spin-polarization direction of the electron beam parallel to the plane of incidence (in-plane$_{\|}$) according to Fig. 2.4d. Figure 2.5c shows the spin-resolved intensity distribution (the SPLEED channeltron intensities N_1 and N_2) of the electron beam with arbitrarily chosen energy E_0. The obtained spin asymmetry $A = SP = 0.071$ is composed of P and the Sherman function S of the detector. With the analysis employed in Budke et al. (2007a) a spin polarization of $P = 29\,\%$ is found, which is in agreement with typical values of GaAs photocathodes.

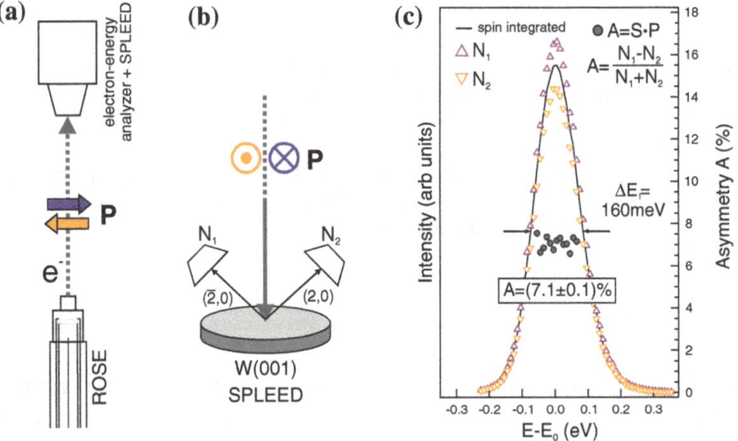

Fig. 2.5 **a** Calibration of the electron beam spin polarization with the SARPES setup. The electron beam from the ROSE is guided to the entrance of the electron-energy analyzer. **b** Subsequent to the analysis of the electron energy, the electron beam is guided onto the W(001) crystal of the SPLEED detector. **c** Spin-resolved intensity distribution (*purple up-* and *orange down-pointing triangles*) and asymmetry A (*gray markers*) of the electron beam measured with the SPLEED detector

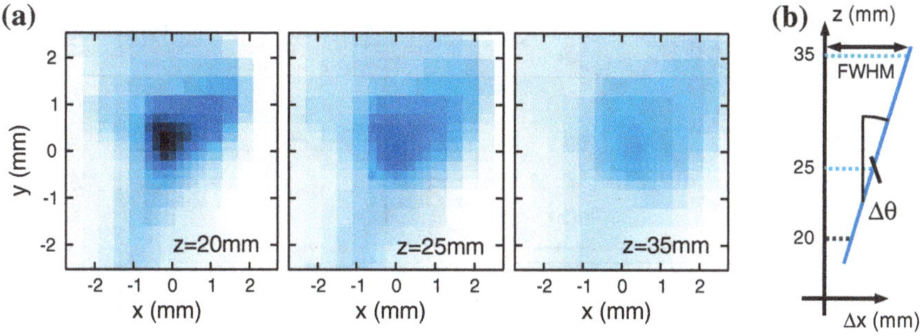

Fig. 2.6 **a** Intensity maps of the electron beam for different positions z behind the last lens element of the ROSE. To determine the angular resolution $\Delta\theta$, the width w (FWHM) of the electron beam is analyzed. The slope a of $w(z)$ relates to the angular resolution $\Delta\theta = \arctan(a)$

(ii) The width of the electron-energy distribution ΔE_{el} is determined by analyzing the width ΔE_I of the intensity distribution from the measurement in Fig. 2.5c. ΔE_I is a convolution of ΔE_{el} with the energy resolution of the electron analyzer ΔE_A: $\Delta E_I^2 = \Delta E_{\text{el}}^2 + \Delta E_A^2$. With $\Delta E_I \approx 160\,\text{meV}$ (FWHM) (see Fig. 2.5c) and $\Delta E_A \approx 95\,\text{meV}$ (FWHM), one obtains $\Delta E_{\text{el}} \approx 129\,\text{meV}$.

(iii) The momentum resolution $\Delta\mathbf{k}$ of the inverse-photoemission experiment solely depends on the beam divergence of the electron beam. In order to get a measure of the angular resolution $\Delta\theta$, the intensity distribution of the electron beam is mapped with a Faraday cup at three different distances z behind the last lens element (see Fig. 2.6a). For each position, the width of the electron beam w (FWHM) is extracted. $\Delta\theta$ is derived as schematically shown in Fig. 2.6b. Deviations from a Gaussian-shaped profile are observed and interpreted as a consequence of astigmatism of the electron transfer optics. The data reveal a beam divergence of $\pm 1.5°$, i.e., a total angular resolution of $\Delta\theta = 3°$, which corresponds to a resolution of the wave vector parallel to the surface at the Fermi level and normal electron incidence of $\Delta k_{\parallel,\text{F}} \approx 0.06\,\text{Å}^{-1}$.

(iv) An ideal electron source would be a point source, followed by an electron optics without lens aberrations. In this case, a truly parallel electron beam can be formed. A real electron source, however, suffers from a finite emission spot (here given by the size of the laser spot), lens aberrations and electron-electron Coulomb repulsion, which becomes increasingly important for low beam energies (here 7–20 eV) and high current densities. Therefore, the beam is optimized with respect to parallelism, i.e., high momentum resolution, at the cost of a small beam size. A beam size d_{el} of less than 2 mm (FWHM) is achieved at the sample located at $z = 25\,\text{mm}$ with beam currents of a few μA (see Fig. 2.6a). For smaller samples, apertures can be used at the last lens element to restrict the beam diameter.

(v) The total energy resolution ΔE of the employed SR-IPE experiment is given by the energy distribution of the electron beam and the width of the bandpass photon detector, the Geiger counter. Experimentally, ΔE is determined by a measurement of the Fermi edge at a polycrystalline Ta sample plate. The energy resolution is derived

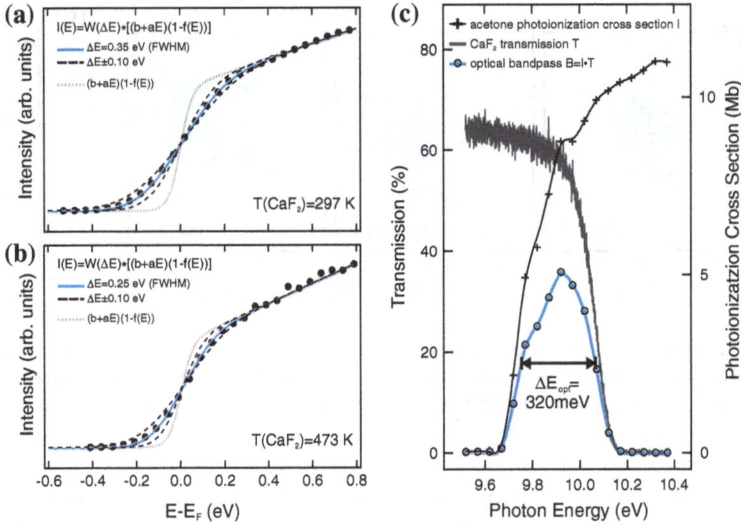

Fig. 2.7 Measurements of the Fermi edge at a polycrystalline Ta sample for **a** $T_{CaF_2} = 297\,\text{K}$ and **b** $T_{CaF_2} = 473\,\text{K}$ of the entrance window of the Geiger counter. *Solid lines* represent simulated spectra $I(E)$ to deduce the total energy resolution ΔE of the SR-IPE experiment. *Dashed lines* show simulated spectra with higher or lower energy resolution to represent confidence intervals. **c** Determination of the optical bandpass with the help of the acetone photoionization cross section I (*black markers* and *line*) from Cool et al. (2005) and the transmission cut-off T of the CaF_2 entrance window (*gray line*) at room temperature. The bandpass (*blue markers* and *line*) is obtained via $B = I \cdot T$ and possesses a width (FWHM) of about $\Delta E_{opt} = 320\,\text{meV}$

by simulating the obtained spectra as discussed in Budke et al. (2007b). It is assumed that the observed spectral intensity $I(E)$ is composed of a constant background b plus a linear slope $a(E)$, multiplied by 1 minus the Fermi function $f(E)$ and convoluted with the respective apparatus function W: $I(E) = W * [(b + aE)(1 - f(E))]$. The full width at half maximum of W represents the energy resolution ΔE of the experiment and is varied to fit the spectra. In the presented setup, to improve the total energy resolution, the optical bandpass of the Geiger counter can be reduced by heating its CaF_2 entrance window (Budke et al. 2007b). Figure 2.7a, b shows the Fermi edge of the polycrystalline Ta for $T_{CaF_2} = 297\,\text{K}$ and $T_{CaF_2} = 473\,\text{K}$, respectively. A total energy resolution of 350 and 250 meV is obtained, respectively, in agreement with values found prior to the applied modifications discussed in Sect. 2.3.1. The optical bandpass is determined by the ionization threshold of the acetone gas (see, e.g., Cool et al. 2005) and the transmission cut-off of the employed CaF_2 window.[2] The multiplication of the threshold with the transmission cut-off represents the optical bandpass (see Fig. 2.7c). At $T_{CaF_2} = 297\,\text{K}$, ΔE_{opt} amounts to $\Delta E_{opt} \approx 320\,\text{meV}$

[2]The transmission cut-off of the CaF_2 window used in this counter was measured in a separate setup (Thiede 2010).

(FWHM). Accordingly, the width of the electron energy distribution is determined to $\Delta E_{el} = (140 \pm 20)$ meV, in agreement with ΔE_{el} found in (ii).

2.4 Performance

Experiments on Ni/W(110) and Au(111) demonstrate the performance of the SR-IPE experiment with the ROSE.

2.4.1 Ni/W(110)

The variable spin sensitivity of the ROSE is demonstrated by means of SR-IPE measurements on the spin-dependent surface electronic structure of a Ni(111) film on W(110). A Ni(111) film with a thickness of 20 monolayers was epitaxially grown on a W(110) single crystal, which had been cleaned by several cycles of heating to 1500 K for 10 min at an oxygen partial pressure of 5×10^{-8} mbar and subsequent flashing to 2300 K. Annealing of the Ni film up to 600 K for 10 min produced a well ordered surface with a sharp diffraction pattern (see Fig. 2.8a). The Ni(111) film is magnetized along the in-plane easy axis $[\bar{1}10]$ ($\bar{\Gamma}\bar{K}$ direction) (Farle et al. 1990) between two coils with a total magnetic field strength of 500 G. This leads to a remanently magnetized film as indicated by the in-situ magneto-optical Kerr-effect measurement (see Fig. 2.8b).

With the state of the film magnetization exactly known, i.e., fixed quantization axis along $\bar{\Gamma}\bar{K}$, the spin-polarization directions of the electron beam can be determined with the help of SR-IPE measurements. The basic approach is as follows. A SR-IPE spectrum is measured at a fixed angle of electron incidence $\theta = 40°$ under four different conditions. For the first two measurements, the ROSE is in the vertical position, i.e., the spin polarization is perpendicular to the plane of incidence. In this

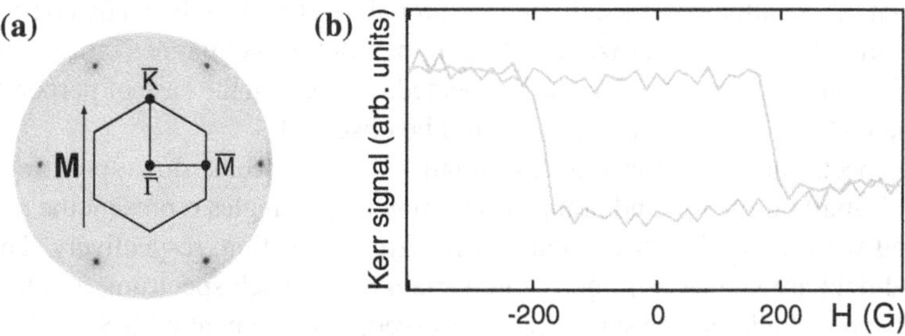

Fig. 2.8 **a** Low-energy electron diffraction image of the Ni(111) film and **b** magneto-optical Kerr-effect measurement (MOKE) with the magnetization direction along $\bar{\Gamma}\bar{K}$

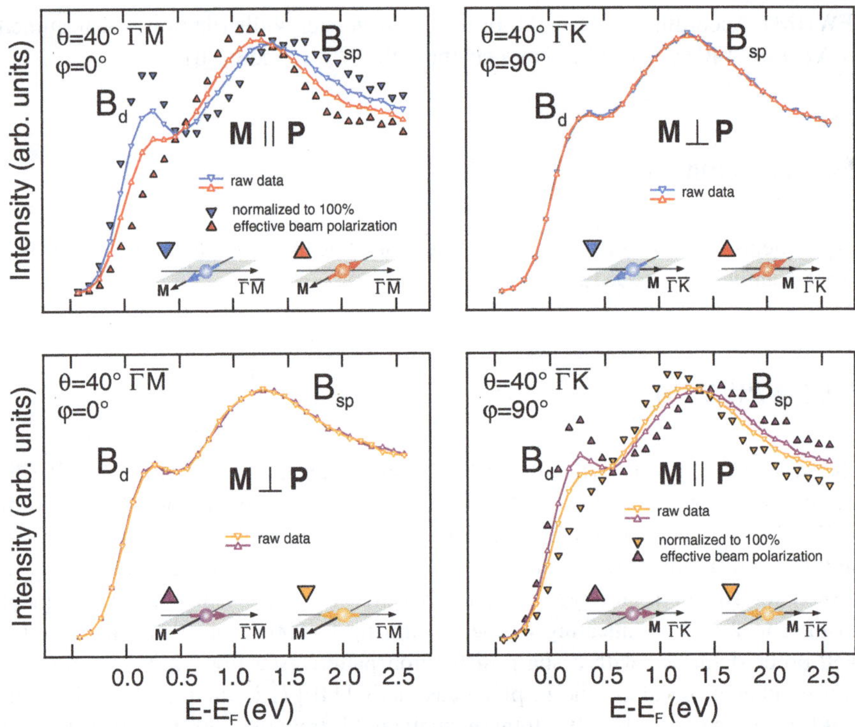

Fig. 2.9 SR-IPE spectra at $\theta = 40°$ on Ni/W(110) magnetized along $\bar{\Gamma}\bar{K}$ with the magnetization direction **M** pointing perpendicular (*left spectra*) or parallel (*right spectra*) to the plane of incidence. Measurements have been conducted with the ROSE in vertical (*upper spectra*) and horizontal setup (*lower spectra*) realizing conditions, where the effective beam polarization $\mathbf{P}_{\text{eff}}$ is parallel or perpendicular to **M**

geometry, spectra are taken with the magnetization direction **M** perpendicular and parallel to the plane of incidence, simply by varying the azimuth φ of the sample. Corresponding measurements are conducted with the ROSE in the horizontal position. Note that in the latter case the effective beam polarization is reduced by cos 40°. For the electron beam spin polarization **P** being parallel to the magnetization direction (**M** ∥ **P**), spin-dependent features are expected, whereas in the case of perpendicular alignment (**M** ⊥ **P**) no polarization should be observed.

The spectra are shown in Fig. 2.9. Small red up- and blue down-pointing triangles and small purple up- and orange down-pointing triangles represent the raw data obtained with the ROSE in vertical and horizontal position, respectively. The normalized data are shown as large colored triangles. In each spectrum, a bulk d-like state B_d and a bulk sp-like state B_{sp} is observed, in agreement with SR-IPE data of Ni(111) bulk single-crystals (Donath 1994).

With the magnetization direction and electron spin polarization pointing perpendicular to the plane of incidence (**M** ∥ **P**, upper left spectrum), clearly, the minority character of B_d and the exchange splitting of B_{sp} are identified. By changing the azimuth of the sample, so that the magnetization direction lies now in the plane of incidence (**M** $\perp$ **P**, upper right spectrum), spin sensitivity is lost. With a rotation of the ROSE to the horizontal position spin sensitivity is regained (**M** $\parallel$ **P**, lower right spectrum) and lost again if the magnetization direction is set to be perpendicular to the plane of incidence (**M** $\perp$ **P**, lower left spectrum). Notably, the spin-integrated spectra along corresponding high-symmetry lines are almost identical. In particular, considering the steep dispersion of B_{sp} (Donath 1994), a rotation of the ROSE, therefore, does not significantly alter the angle of electron incidence. Also the sample condition is not affected. Evidently, the ROSE gives access to two orthogonal in-plane spin-polarization directions, the in-plane$_\perp$ and the in-plane$_\parallel$ spin-polarization direction.

2.4.2 Au(111)

The study of the Au(111) surface highlights the importance of a correct electron beam and sample alignment, when investigating samples with spin textures that are more complex than in ferromagnetic systems, such as Ni/W(110), with a single spin quantization axis along the magnetization direction.

The surface electronic structure of Au(111) is discussed in several theoretical and experimental studies and is the prototypical Rashba system (Paniago et al. 1995; LaShell et al. 1996; Nicolay et al. 2001; Henk et al. 2003; Reinert 2003; Hoesch et al. 2004; Henk et al. 2004a, b). It features a nearly-free-electron-like dispersing spin-orbit-split Shockley surface state, which exhibits a spin texture as predicted by the Rashba-Bychkov model. Recently, SR-IPE investigations confirmed the Rashba-type spin texture of the unoccupied part of the Au(111) surface state (Wissing et al. 2013). Deviations from a nearly-free-electron-like dispersion occur as soon as the surface state approaches the gap boundary and becomes a surface-resonant state. Figure 2.10a shows a sketch of the $E(\mathbf{k}_\parallel)$ dispersion of the Au(111) surface state in the vicinity of $\bar{\Gamma}$, where only minor deviations from a parabolic dispersion occur.

As a first step, the spin-orbit-split surface state is identified with the help of SR-IPE measurements for three angles of electron incidence θ along $\bar{\Gamma}\bar{M}$ as illustrated in Fig. 2.10a: $\theta = 14°, 7°$ and $0°$. Prior to the experiment, the Au(111) surface was cleaned by several cycles of sputtering and annealing, whereas the cleanliness and crystalline quality of the surface were checked with AES, LEED, and STM. Figure 2.10b presents the SR-IPE data. The angle of electron incidence θ has been calibrated with the help of additional SR-IPE measurements (see Fig. 2.10c). The sensitivity of the experiment is set to the in-plane$_\perp$ spin-polarization direction as denoted by the red up- and blue down-pointing triangles. Spectra have been normalized to 100 % in-plane$_\perp$ spin polarization of the electron beam.

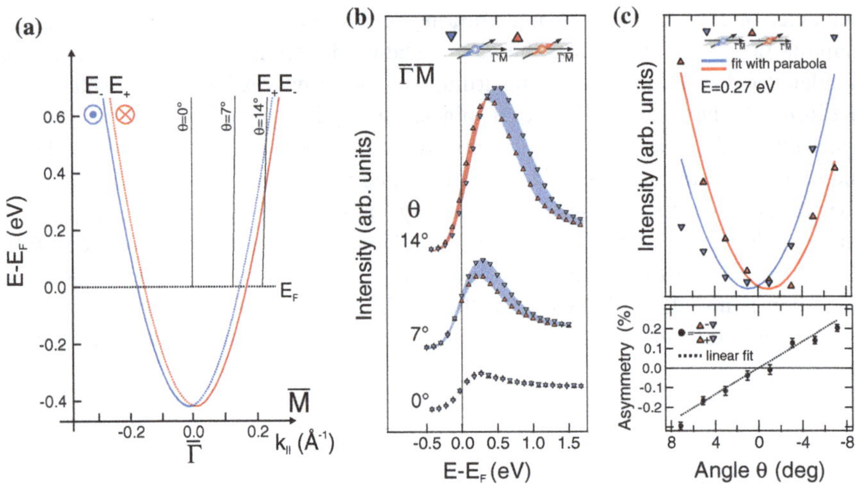

Fig. 2.10 **a** Sketch of the Au(111) Shockley surface state in the vicinity of the $\bar{\Gamma}$ point along $\bar{\Gamma}\bar{\mathrm{M}}$. The spin-polarization direction of the two spin-orbit-split surface-state components is in-plane and perpendicular to $\mathbf{k}_{\parallel}$ (in-plane$_{\perp}$). SR-IPE measurements at constant angles of electron incidence θ scan the surface state as illustrated by the *dotted lines*. **b** SR-IPE measurements of the Au(111) surface with sensitivity to the in-plane$_{\perp}$ spin-polarization direction along $\bar{\Gamma}\bar{\mathrm{M}}$ for $\theta = 0°, 7°$ and $14°$. **c** (*upper graph*) Spin-resolved intensity at $E = 0.27\,\mathrm{eV}$ for different angles of electron incidence θ and (*lower graph*) according spin-asymmetry data

(i) At $\theta = 14°$, both spin-orbit-split surface-state components E_+ and E_- are completely unoccupied. These can be clearly observed in the SR-IPE spectra. The energy splitting is approximately $\Delta E \approx 150\,\mathrm{meV}$.

(ii) The spin-orbit-split surface-state components become unoccupied for momenta parallel to the surface of about $k_{\parallel,+} \approx 0.15\,\text{Å}^{-1}$ and $k_{\parallel,-} \approx 0.18\,\text{Å}^{-1}$, respectively. At $\theta = 7°$ ($k_{\parallel,+,-} \approx 0.13\,\text{Å}^{-1}$), both surface-state components are located below E_{F}. However, spectral intensity close to E_{F} is observed. This is a consequence of the limited momentum resolution in combination with the Fermi-level cut-off and the finite lifetime of the surface state of the SR-IPE experiment (Wissing et al. 2013). In addition, the observed peak positions close to E_{F} are influenced by the limited energy resolution and do not necessarily represent the final-state energies (Donath 1989; Budke et al. 2007a; Stolwijk et al. 2010). The observed intensity difference between both spin channels is owed to the fact that one spin component is located closer to E_{F} than the other.

(iii) At $\theta = 0°$ ($\bar{\Gamma}$), the surface-state components are degenerate and completely occupied. In the SR-IPE spectra, a small spectral intensity close to the Fermi level remains. In agreement with the Rashba-Bychkov model, no in-plane$_{\perp}$ spin polarization is detected. Note that the observation of spin asymmetry close to the Fermi level is highly sensitive to slight variations of θ, e.g., for $-2°$ and $3°$ intensity asymmetries are detected, due to the same reasons as discussed for $\theta = 7°$ (Wissing et al. 2013). In return, this can be used to calibrate θ. Figure 2.10c

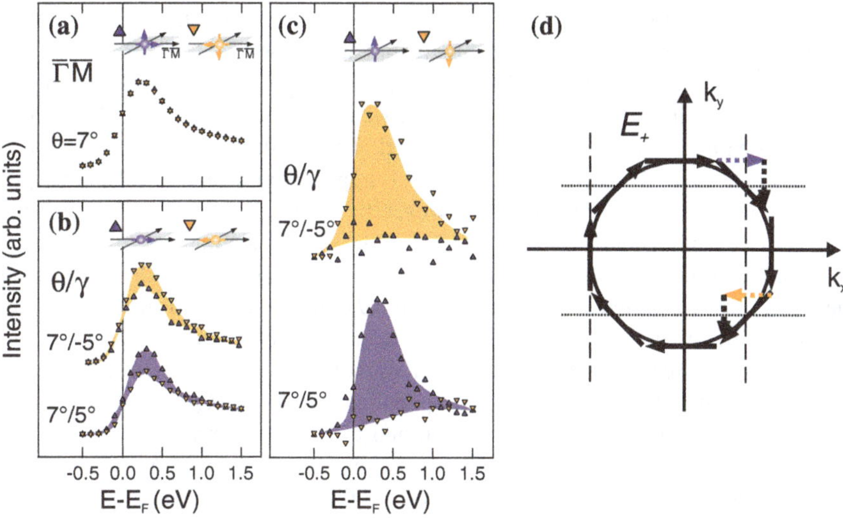

Fig. 2.11 **a** SR-IPE measurement for $\theta = 7°$ with sensitivity to the out-of-plane/in-plane$_{\parallel}$ spin-polarization direction. **b, c** The same measurement as in (**a**) with intentionally tilted electron beam by $\gamma = \pm 5°$. Spectra in (**b**) and (**c**) are obtained from the same raw data. In (**b**), a normalization to 100 % in-plane$_{\parallel}$ beam polarization is applied, in (**c**), the raw data are normalized to 100 % out-of-plane beam polarization. **d** Illustration of the dispersion and spin-polarization direction (*arrows*) of the Au(111) surface-state component E_+ in the k_x-k_y plane. The *dashed* (*dotted*) *lines* correspond to measurements with $k_x \neq 0°$ ($k_y \neq 0°$). For $k_x \neq 0°$ and $k_y \neq 0°$ (crossings of the *dashed* and *dotted lines*), the spin-polarization vector features components along k_x (*orange* and *purple dotted arrows*) and k_y (*black dotted arrows*)

presents a measurement of the spin-resolved intensity at $E = 0.27$ eV as a function of θ. The spin-asymmetry data ($A = (I_\uparrow - I_\downarrow)/(I_\uparrow + I_\downarrow)$) illustrate the reversal of the spin-polarization direction between positive and negative angles. Spin asymmetry is only absent at $\theta = 0°$.

Clearly, the measurements detect the Au(111) spin-orbit-split Shockley surface state. As predicted by the Rashba-Bychkov model the states are in-plane$_\perp$ spin polarized.

The $\bar{\Gamma}\bar{\text{M}}$ direction lies in the mirror plane of the system and out-of-plane spin polarization is neither expected theoretically (Henk et al. 2004b) nor measured by SARPES experiments (Hoesch et al. 2004). In Fig. 2.11a, the SR-IPE measurement for $\theta = 7°$ along $\bar{\Gamma}\bar{\text{M}}$ with the spin sensitivity set to the out-of-plane/in-plane$_{\parallel}$ spin-polarization direction (purple up- and orange down-pointing triangles) is able to confirm this finding. Intriguingly, spin asymmetries are observed, when the electron beam is intentionally tilted by γ along the direction orthogonal to $\bar{\Gamma}\bar{\text{M}}$. Figure 2.11b presents SR-IPE spectra taken at $\theta = 7°$ with a tilt of about $\gamma = 5°$ and $\gamma = -5°$. Obviously, the spin quantization axis of the surface state has changed so that out-of-plane and/or in-plane$_{\parallel}$ spin-polarization components emerge. This would be impossible for systems with fixed spin quantization axis such as ferromagnetic surfaces. Note that a normalization to 100 % in-plane$_{\parallel}$ beam polarization has been applied, without knowing the exact spin quantization axis.

By way of trial, it is assumed that the observed spin asymmetry is a result of a pure out-of-plane spin polarization. In this case, a normalization to 100 % out-of-plane beam polarization has to be applied. The resulting spectra are shown in Fig. 2.11c. The SR-IPE measurements would suggest completely out-of-plane spin-polarized surface-state components. However, theory predicts an out-of-plane spin-polarization component of below 5 % (Henk et al. 2004b). An interpretation in terms of large out-of-plane spin-polarization components is, therefore, unlikely. Instead, the results are interpreted as follows. A rotation θ of the sample leads to a momentum $\mathbf{k}_x$ along $\bar{\Gamma}\bar{\mathrm{M}}$. A tilt γ effectively produces a momentum $\mathbf{k}_y$ orthogonal to $\mathbf{k}_x$ and parallel to $\bar{\Gamma}\bar{\mathrm{K}}$. According to the Rashba-Bychkov model, the quantization axis of the surface-state spin polarization is perpendicular to $\mathbf{k}_{\parallel} = (k_x, k_y)$. A rotation θ, thus, leads to in-plane$_{\perp}$ spin polarization and a tilt γ gives rise to in-plane$_{\parallel}$ spin polarization. Figure 2.11d illustrates the surface-state component E_+ in the k_x-k_y plane. For $\theta \neq 0°$ ($k_x \neq 0$) and $\gamma \neq 0°$ ($k_y \neq 0$), E_+ exhibits a spin-polarization component parallel to $\mathbf{k}_x$. The value of the in-plane$_{\parallel}$ spin polarization P_x depends on the ratio $\frac{k_y}{|k_{\parallel}|}$, see, e.g., Bentmann et al. (2011). In the case of the Au(111) surface state, for $\theta = 7°$ and $\gamma = 5°$, $\mathbf{P}_x$ is large enough to be detected. The spin polarization reversal between $\gamma = 5°$ and $\gamma = -5°$ affirms this argumentation.

Ultimately, the strength of the SR-IPE experiment with the ROSE is to analyze unoccupied states with respect to more than just one spin-polarization direction and, thereby, uncover complex spin textures. The Au(111) spin-orbit-split Shockley surface state is identified. It is demonstrated that the spin texture of the surface-state components is in agreement with the Rashba-Bychkov model. Along $\bar{\Gamma}\bar{\mathrm{M}}$, only in-plane$_{\perp}$ spin polarization is detected. Notably, it is shown that in systems, where the quantization axis of the spin polarization is more complex than in single axis systems, a correct sample and electron beam alignment is indispensable. For the Au(111) surface state, a tilt of the electron beam of $\gamma = 5°$ already leads to an observation of in-plane$_{\parallel}$ spin polarization, which can be easily misinterpreted as out-of-plane spin polarization, if the alignment is not exactly known.

References

Allmers, T.: Magnetic nanostructures: influence of morphology on electronic and magnetic properties, Ph.D. thesis, Münster University (2009)

Allmers, T., Donath, M.: Growth and morphology of thin Fe films on flat and vicinal Au(111): a comparative study. New J. Phys. **11**, 103049 (2009)

Allmers, T., Donath, M.: Magnetic properties of Fe films on flat and vicinal Au(111): consequences of different growth behavior. Phys. Rev. B **81**, 064405 (2010)

Allmers, T., Donath, M., Braun, J., Minár, J., Ebert, H.: d- and sp-like surface states on fcc Co(001) with distinct sensitivity to surface roughness. Phys. Rev. B **84**, 245426 (2011)

Auger, P.: Sur l'effet photoélectrique composé. J. Phys. Radium **6**, 205 (1925)

Bentmann, H., Kuzumaki, T., Bihlmayer, G., Blügel, S., Chulkov, E.V., Reinert, F., Sakamoto, K.: Spin orientation and sign of the Rashba splitting in Bi/Cu(111). Phys. Rev. B **84**, 115426 (2011)

Binnig, G., Rohrer, H.: Scanning tunneling microscopy. IBM J. Res. Dev. **44**, 279 (2000)

Budke, M., Allmers, T., Donath, M., Rangelov, G.: Combined experimental setup for spin- and angle-resolved direct and inverse photoemission. Rev. Sci. Instrum. **78**, 113909 (2007a)

Budke, M., Donath, M.: Ar gas discharge lamp with heated LiF window: a monochromatized light source for photoemission. Appl. Phys. Lett. **92**, 231918 (2008)

Budke, M., Donath, M.: Influence of impurities on the electronic structure of Y(0001) single-crystal surfaces. Phys. Rev. B. **79**, 075432 (2009)

Budke, M., Renken, V., Liebl, H., Rangelov, G., Donath, M.: Inverse photoemission with energy resolution better than 200 meV. Rev. Sci. Instrum. **78**, 083903 (2007b)

Burnett, G.C., Monroe, T.J., Dunning, F.B.: High-efficiency retarding-potential Mott polarization analyzer. Rev. Sci. Instrum. **65**, 1893 (1994)

Chen, C.J.: Introduction to Scanning Tunneling Microscopy. Oxford University Press, New York (2008)

Cool, T.A., Wang, J., Nakajima, K., Taatjes, C.A., McIlroy, A.: Photoionization cross sections for reaction intermediates in hydrocarbon combustion. Int. J. Mass Spectrom. **247**, 18 (2005)

Desinger, K., Dose, V., Göbl, M., Scheidt, H.: Momentum resolved bremsstrahlung isochromat spectra from Ni(001). Solid State Commun. **49**, 479 (1984)

Donath, M.: Spin-resolved inverse photoemission of ferromagnetic surfaces. Appl. Phys. A **49**, 351 (1989)

Donath, M.: Spin-dependent electronic structure at magnetic surfaces: the low-Miller-index surfaces of nickel. Surf. Sci. Rep. **20**, 251 (1994)

Donath, M.: Magnetic order and electronic structure in thin films. J. Phys.: Condens. Matter **11**, 9421 (1999)

Donath, M., Glöbl, M., Senftinger, B., Dose, V.: Photon polarization effects in inverse photoemission from Cu(001). Solid State Commun **60**, 237 (1986)

Dose, V.: VUV isochromat spectroscopy. Appl. Phys. B **14**, 117 (1977)

Dose, V.: Momentum-resolved inverse photoemission. Surf. Sci. Rep. **5**, 337 (1985)

Duden, T., Bauer, E.: A compact electron-spin-polarization manipulator. Rev. Sci. Instrum. **66**, 2861 (1995)

Eberhardt, W., Himpsel, F.J.: Dipole selection rules for optical transitions in the fcc and bcc lattices. Phys. Rev. B **21**, 5572 (1980)

Escher, M., Weber, N.B., Merkel, M., Plucinski, L., Schneider, C.M.: FERRUM: a new highly efficient spin detector for electron spectroscopy. e-J. Surf. Sci. Nanotech. **9**, 340 (2011)

Farle, M., Berghaus, A., Li, Y., Baberschke, K.: Hard-axis magnetization of ultrathin Ni(111) films on W(110): an experimental method to measure the magneto-optic Kerr effect in ultrahigh vacuum. Phys. Rev. B **42**, 4873 (1990)

Fauster, T., Schneider, R., Dürr, H.: Angular distribution of the inverse photoemission from Cu(100). Phys. Rev. B **40**, 7981 (1989)

Heinzmann, U., Dil, J.H.: Spin-orbit-induced photoelectron spin polarization in angle-resolved photoemission from both atomic and condensed matter targets. J. Phys.: Condens. Matter **24**, 173001 (2012)

Henk, J., Ernst, A., Bruno, P.: Spin polarization of the L-gap surface states on Au(111). Phys. Rev. B **68**, 165416 (2003)

Henk, J., Ernst, A., Bruno, P.: Spin polarization of the L-gap surface states on Au(111): a first-principles investigation. Surf. Sci. **566**, 482 (2004a)

Henk, J., Hoesch, M., Osterwalder, J., Ernst, A., Bruno, P.: Spin-orbit coupling in the L-gap surface states of Au(111): spin-resolved photoemission experiments and first-principles calculations. J. Phys.: Condens. Matter **16**, 7581 (2004b)

Henk, J., Scheunemann, T., Halilov, S.V., Feder, R.: Magnetic dichroism and electron spin polarization in photoemission: analytical results. J. Phys.: Condens. Matter **8**, 47 (1996)

Henzler, M., Goepel, W.: Introduction to Solid State Physics, Teubner Studienbücher (1994)

Himpsel, F.J.: Inverse photoemission from semiconductors. Surf. Sci. Rep. **12**, 3 (1990)
Hoesch, M., Greber, T., Petrov, V., Muntwiler, M., Hengsberger, M., Auwärter, W., Osterwalder, J.: Spin-polarized Fermi surface mapping. J. Electron Spectrosc. Relat. Phenom. **124**, 263 (2002)
Hoesch, M., Muntwiler, M., Petrov, V.N., Hengsberger, M., Patthey, L., Shi, M., Falub, M., Greber, T., Osterwalder, J.: Spin structure of the Shockley surface state on Au(111). Phys. Rev. B **69**, 241401 (2004)
Johnson, P.D.: Spin-polarized photoemission. Rep. Prog. Phys. **60**, 1217 (1997)
Kerr, J.: On rotation of the plane of the polarization by reflection from the pole of a magnet. Philos. Mag. **3**, 321 (1877)
Kirschner, J., Feder, R.: Spin polarization in double diffraction of low-energy electrons from W(001): experiment and theory. Phys. Rev. Lett. **42**, 1008 (1979)
Kittel, C.: Introduction to Solid State Physics. John Wiley, New York (1967)
Kolac, U., Donath, M., Ertl, K., Liebl, H., Dose, V.: High-performance GaAs polarized electron source for use in inverse photoemission spectroscopy. Rev. Sci. Instrum. **59**, 1933 (1988)
Kopitzki, K.: Einführung in die Festkörperphysik, Teubner Studienbücher (1989)
LaShell, S., McDougall, B.A., Jensen, E.: Spin splitting of an Au(111) surface state band observed with angle resolved photoelectron spectroscopy. Phys. Rev. Lett. **77**, 3419 (1996)
Linsmeier, C.: Auger electron spectroscopy. Vacuum **45**, 673 (1994)
Nicolay, G., Reinert, F., Hüfner, S., Blaha, P.: Spin-orbit splitting of the L-gap surface state on Au(111) and Ag(111). Phys. Rev. B **65**, 033407 (2001)
Okuda, T., Kimura, A.: Spin- and angle-resolved photoemission of strongly spin-orbit coupled systems. J. Phys. Soc. Jpn. **82**, 021002 (2013)
Okuda, T., Miyamaoto, K., Miyahara, H., Kuroda, K., Kimura, A., Namatame, H., Taniguchi, M.: Efficient spin resolved spectroscopy observation machine at Hiroshima Synchrotron Radiation Center. Rev. Sci. Instrum. **82**, 103302 (2011)
Okuda, T., Takeichi, Y., Maeda, Y., Harasawa, A., Matsuda, I., Kinoshita, T., Kakizaki, A.: A new spin-polarized photoemission spectrometer with very high efficiency and energy resolution. Rev. Sci. Instrum. **79**, 123117 (2008)
Paniago, R., Matzdorf, R., Meister, G., Goldmann, A.: Temperature dependence of Shockley-type surface energy bands on Cu(111), Ag(111) and Au(111). Surf. Sci. **336**, 113 (1995)
Pendry, J.B.: New probe for unoccupied bands at surfaces. Phys. Rev. Lett. **45**, 1356 (1980)
Peng, L., Dudarev, S., Whelan, M.: High-Energy Electron Diffraction and Microscopy, Oxford University Press, New York (2011)
Pierce, D.T., Celotta, R.J., Wang, G.C., Unertl, W.N., Galejs, A., Kuyatt, C.E., Mielczarek, S.R.: The GaAs spin polarized electron source. Rev. Sci. Instrum. **51**, 478 (1980)
Pierce, D.T., Meier, F.: Photoemission of spin-polarized electrons from GaAs. Phys. Rev. B **13**, 5484 (1976)
Polisetty, S., Scheffler, J., Sahoo, S., Wang, Y., Mukherjee, T., He, X., Binek, C.: Optimization of magneto-optical Kerr setup: analyzing experimental assemblies using Jones matrix formalism. Rev. Sci. Instrum. **79**, 055107 (2008)
Powell, C.: The quest for universal curves to describe the surface sensitivity of electron spectroscopies. J. Electron Spectrosc. Relat. Phenom. **47**, 197 (1988)
Rangelov, G., Kang, H.D., Reinmuth, J., Donath, M.: Surface magnetic properties of ultrathin Fe films on W(110) studied by spin-resolved appearance potential spectroscopy. Phys. Rev. B **61**, 549 (2000)
Reinert, F.: Spin-orbit interaction in the photoemission spectra of noble metal surface states. J. Phys.: Condens. Matter **15**, S693 (2003)
Schönhense, G., Heinzmann, U.: A capillary discharge tube for the production of intense VUV resonance radiation. J. Phys. E: Sci. Instrum. **16**, 74 (1983)
Smith, N.V.: Inverse photoemission. Rep. Prog. Phys. **51**, 1227 (1988)
Stolwijk, S.D.: Die elektronische Struktur von Yttrium-(0001), Diploma thesis, Münster University (2009)

Stolwijk, S.D., Schmidt, A.B., Donath, M.: Surface state with d_{z^2} symmetry at Y(0001): a combined direct and inverse photoemission study. Phys. Rev. B **82**, 201412(R) (2010)

Stolwijk, S.D., Wortelen, H., Schmidt, A.B., Donath, M.: Rotatable spin-polarized electron source for inverse-photoemission experiments. Rev. Sci. Instrum. **85**, 013306 (2014)

Thiede, C.: Über Transmissionsmessungen an CaF_2-Kristallen im Vakuum-ultravioletten Spektralbereich, Diploma thesis, Münster University (2010)

Winkelmann, A., Hartung, D., Engelhard, H., Chiang, C.T., Kirschner, J.: High efficiency electron spin polarization analyzer based on exchange scattering at Fe/W(001). Rev. Sci. Instrum. **79**, 083303 (2008)

Wissing, S.N.P., Eibl, C., Zumbülte, A., Schmidt, A.B., Braun, J., Minár, J., Ebert, H., Donath, M.: Rashba-type spin splitting at Au(111) beyond the Fermi level: the other part of the story. New J. Phys. **15**, 105001 (2013)

Yu, D., Math, C., Meier, M., Escher, M., Rangelov, G., Donath, M.: Characterisation and application of a SPLEED-based spin polarisation analyser. Surf. Sci. **601**, 5803 (2007)

Chapter 3
Spin Textures on Tl/Si(111)-(1×1)

3.1 Motivation

The main topic of this thesis addresses the spin-dependent surface electronic structure of Tl/Si(111)-(1×1). The relevance of Tl/Si(111)-(1×1) is twofold.

(i) The development of spintronics, i.e., the use of the electron spin as information carrier in electronics, hinges on the generation and manipulation of spin-polarized currents. An overview on the large scope of spintronics is found, e.g., in Žutić et al. (2004), Fabian et al. (2007), Awschalom and Flatte (2007) and Dyakonov (2008). Here, the Rashba-Bychkov effect (Bychkov and Rashba 1984), i.e., the lifting of the spin degeneracy due to spin-orbit interaction and space inversion asymmetry, opens the way for promising applications such as spin field effect transistors (Datta and Das 1990; Schliemann et al. 2003) or spin Hall effect transistors (Wunderlich et al. 2010). In recent years, it has been demonstrated for several metal surfaces that the Rashba-Bychkov effect leads to a spin-dependent splitting of surface states (see, e.g., LaShell et al. 1996; Hochstrasser et al. 2002; Hoesch et al. 2004; Koroteev et al. 2004; Sugawara et al. 2006; Tamai et al. 2013). To implement the surface Rashba-Bychkov effect in spintronics one has to satisfy three requirements: (i) The surface state has to be metallic to contribute to the current. (ii) The spin splitting of the surface state should be larger than the lifetime broadening, so that one is able to distinguish between the two spin components even at room temperature. A large splitting would also decrease the spin-precession time in a spin transistor (Nitta et al. 1997). In some cases, such as Bi/Ag(111) (Ast et al. 2007) and Bi/Si(111) (Gierz et al. 2009), giant splittings of about 0.2 eV, orders of magnitude larger than those found in semiconductor heterostructures (Winkler 2003), could be observed.

Figures and text excerpts of this chapter have been published in Stolwijk et al. (2013). Reprinted with permission from Stolwijk et al. (2013). Copyright (2013) by the American Physical Society.

S.D. Stolwijk, *Spin-Orbit-Induced Spin Textures of Unoccupied Surface States on Tl/Si(111)*, Springer Theses, DOI 10.1007/978-3-319-18762-4_3

(iii) The substrate has to be semiconducting, so that only the surface state contributes to the spin transport.
Thin films of heavy metals on semiconducting substrates are likely to possess all necessary ingredients and lie in the focus of recent research (Barke et al. 2006; Sakamoto et al. 2009; Gierz et al. 2009; Hatta et al. 2009; Yaji et al. 2010; Mathias et al. 2010; Okuda et al. 2010; Takayama et al. 2011; Kocán et al. 2011; Ohtsubo et al. 2012; Höpfner et al. 2012; Bondarenko et al. 2013; Sakamoto et al. 2013; Stolwijk et al. 2013). In this context, the Tl (1×1) adlayer on Si(111) poses an excellent candidate.

(ii) In recent years, increasingly complex spin-orbit-induced spin textures in momentum space have been theoretically predicted and experimentally detected for various surfaces. This includes the occurrence of out-of-plane spin polarization in the Dirac cones of topological insulators (Souma et al. 2011; Basak et al. 2011; Herdt et al. 2013; Nomura et al. 2014) and rotations of spin-polarization vectors of spin-orbit-split surface states in momentum space (Sakamoto et al. 2009; Ibañez-Azpiroz et al. 2011; Takayama et al. 2011; Höpfner et al. 2012; Ohtsubo et al. 2012; Sakamoto et al. 2013; Stolwijk et al. 2013). Tl/Si(111)-(1×1) features an unoccupied surface state, which can be traced throughout the complete Brillouin zone. In this light, the Tl/Si(111)-(1×1) surface is the prototypical example to study the interplay between spin-orbit interaction and the surface symmetry. The threefold honeycomb layered structure of the Tl/Si(111)-(1×1) surface in combination with the strong spin-orbit coupling of the adlayer leads to complex spin textures in momentum space, which can not be described by the Rashba-Bychkov model.

Outline

The outline of this chapter is as follows: Sect. 3.2 is concerned with the preparation and characterization of the sample system Tl/Si(111)-(1×1). Here, particular attention is paid to the symmetry of the Tl/Si(111)-(1×1) surface. The results on the spin-dependent surface electronic structure are presented in Sects. 3.3–3.7.

Section 3.3 addresses the $\bar{\Gamma}\bar{K}$ ($\bar{\Gamma}\bar{K}'$) high-symmetry directions. Theoretical predictions and experimental results for the occupied surface electronic structure set the stage for the study on the unoccupied surface electronic structure. Spin- and angle-resolved inverse-photoemission measurements with the ROSE, i.e., with sensitivity to both the in-plane and the out-of-plane spin-polarization direction, identify an unoccupied spin-orbit-split surface state with unique properties, which are well described by theoretical calculations. As will be demonstrated, this surface state extends over the complete surface Brillouin zone. The behavior of the surface state will be evaluated with respect to the surface quality, adsorbate exposure and $E(\mathbf{k}_\perp)$ dispersion. Furthermore, it is demonstrated that the spin-polarization vector rotates from the in-plane$_\perp$ spin-polarization direction around $\bar{\Gamma}$ to the direction perpendicular to the surface at the $\bar{K}$ ($\bar{K}'$) points—a direct consequence of the symmetry of the 2D hexagonal system.

In Sect. 3.4, the focus is put on the $\bar{K}$ ($\bar{K}'$) points. Here, the unoccupied surface-state components form spin-orbit-split valleys. A giant splitting in energy of about 0.6 eV is detected, which is attributed to the strong localization of the unoccupied surface state close to the heavy Tl atoms. This leads to completely out-of-plane spin-polarized valleys in the vicinity of the Fermi level. As the valley polarization is oppositely oriented at the $\bar{K}$ and $\bar{K}'$ points, backscattering is strongly suppressed in this system. Moreover, it is shown that doping of the Tl/Si(111)-(1×1) surface by adsorption of additional Tl gives rise to metallic spin-polarized valleys, making the Tl/Si(111)-(1×1) surface highly interesting for spintronic and valleytronic applications.

As part of Sect. 3.5, it will be demonstrated that the unoccupied surface state becomes a surface-resonant state along the $\bar{\Gamma}\bar{M}$ directions. It is shown that the surface-resonant state exhibits a giant spin splitting around $\bar{M}$, which is in the same order of magnitude as the giant splitting found for the surface alloy Bi/Ag(111) (Ast et al. 2007). Furthermore, the SR-IPE experiments verify that the surface resonance is purely in-plane$_\perp$ spin polarized as demanded by symmetry arguments. Notably, right next to the high-symmetry line this constraint is lifted and additional spin-polarization directions emerge.

In Sect. 3.6, the $\bar{K}\bar{M}$ ($\bar{K}'\bar{M}$) high-symmetry directions are investigated. Here, an unoccupied surface state is detected, which can be seen as the continuation of the unoccupied surface state observed along $\bar{\Gamma}\bar{K}$. Similar to the $\bar{\Gamma}\bar{M}$ direction, the surface state exhibits a giant spin splitting around the $\bar{M}$ point, which even exceeds the giant splitting observed on Bi/Ag(111) (Ast et al. 2007). The spin texture of the surface state features out-of-plane and in-plane$_\perp$ spin-polarization components. However, it will be demonstrated that out-of-plane spin polarization is dominant.

The combined results for the $\bar{\Gamma}\bar{M}$ and $\bar{K}\bar{M}$ ($\bar{K}'\bar{M}$) directions indicate a peculiar spin texture around $\bar{M}$. Light is shed onto the spin texture by fitting an effective Hamiltonian to the experimental results. At a given energy, the $\bar{M}$ point is encircled by a surface-state component, which is almost completely out-of-plane spin polarized at the $\bar{K}\bar{M}$ ($\bar{K}'\bar{M}$) intersections and purely in-plane spin polarized at the $\bar{\Gamma}\bar{M}$ intersections. The resulting spin texture is interpreted as a spin chirality in momentum space.

General Remarks

It is remarked that all SR-IPE measurements were conducted with the sample kept at room temperature. Cooling to 120 K had no influence on the obtained spectra concerning the energetic positions and spectral intensities of the observed features. Furthermore, the relevant state transitions measured in this work could be most clearly observed with counter $C1$ ($\Delta E = 350$ meV), while measurements with improved energy resolution with counter $C2$ and $C4$ ($\Delta E = 250$ meV) did not lead to significantly improved spectra (see Fig. 3.9). Therefore, unless otherwise noted, SR-IPE data shown in this thesis were detected with counter $C1$. The angle of electron incidence θ and the azimuth of the sample φ were calibrated with the help of SR-IPE measurements, e.g., by monitoring the spin-dependent photon yield at a given energy for varying θ or φ (see, e.g., Figs. 3.16 and 3.21).

3.2 The Sample System

Along with aluminum (Al), gallium (Ga) and indium (In), Tl belongs to the metals of the third (III) group of the periodic table. With an atomic number of $Z = 81$, Tl neighbors mercury (Hg, $Z = 80$), lead (Pb, $Z = 82$) and bismuth (Bi, $Z = 83$). Tl has an electronic configuration of [Xe]$4f^{14}5d^{10}6s^{2}6p^{1}$ and crystallizes in the hexagonal closed-packed (hcp) structure. Tl is chemically peculiar, in that it can occur in a monovalent oxidation state, in addition to the trivalent state, which is encountered for all the other group III elements (Greenwood and Earnshaw 1984). In the monovalent oxidation state, the two s electrons behave as inert pair, which is referred to as *inert pair effect*.

The adsorption of thallium on Si(111)-(7×7) leads to a number of interesting surface phases, which are discussed vividly in literature in relation to the influence of the *inert pair effect* (Vitali et al. 1999b, 2001; Kotlyar et al. 2002, 2003; Özkaya et al. 2008; Kocán et al. 2011, 2012). Similar to structures found for the other group III metals (Lai and Wang 2001; Li et al. 2002), Tl forms a superlattice of metallic nanoclusters (magic clusters) for coverages of about 0.2 monolayer (ML) (1 ML= 7.8×10^{14} cm^{-2}) (Vitali et al. 1999b; Kotlyar et al. 2002). For 1/3 ML, Tl is found to form a mosaic ($\sqrt{3} \times \sqrt{3}$) structure (Kotlyar et al. 2003; Kocán et al. 2012), which resembles but is not alike the ($\sqrt{3} \times \sqrt{3}$) structure of Al, Ga and In (Nicholls et al. 1987; Zegenhagen et al. 1989; Takaoka et al. 1993; Jia et al. 2002; Mizuno et al. 2003). This observation is in agreement with photoemission results, which find a monovalent state for Tl in the Tl/Si(111)-($\sqrt{3} \times \sqrt{3}$) structure (Sakamoto et al. 2006, 2007). Adsorption of more than one ML up to 50 ML Tl leads to the formation of few but large Tl islands (Vitali et al. 2001).

In contrast to the other group III elements, the adsorption of 1 ML Tl on Si(111) gives rise to a pseudomorphic (1×1)-Tl adlayer as shown in Fig. 3.2a. Studies on the Tl/Si(111)-(1×1) surface with STM (Vitali et al. 1999a; Kotlyar et al. 2003), LEED (Noda et al. 2003), synchrotron x-ray scattering (Kim et al. 2004) and ab-initio calculations (Lee et al. 2002; Stolwijk et al. 2013) show that the Tl atoms occupy the T_4 adsorption site, which is anomalous among metals deposited on Si(111) (Noda et al. 2003). In particular, no (1×1) monolayer structures on Si(111) are found for other heavy metals like Bi, Pb and Au (Wan et al. 1992; Cheng and Kunc 1997; Nagao et al. 1998; Kirchmann et al. 2007; Švec et al. 2008; Bondarenko et al. 2013). The perfect Tl/Si(111)-(1×1) surface is semiconducting (Lee et al. 2002), whereas Tl multivacancies may lead to metallicity (Kocán et al. 2011).

Please note that one monolayer of Tl on Ge(111) results in a Tl/Ge(111)-(1×1) surface, which is isoelectronic to Tl/Si(111)-(1×1) with respect to the valence electrons. The Tl/Ge(111)-(1×1) surface is studied in several publications with LEED (Hatta et al. 2007), STM (Castellarin-Cudia et al. 2001) and ARPES (Hatta et al. 2007; Ohtsubo et al. 2012). Band structure calculations show strong similarities to Tl/Si(111)-(1×1) (Hatta et al. 2007; Ohtsubo et al. 2012).

In the following, the preparation of well-ordered Tl/Si(111)-(1×1) surfaces will be described in two steps. First, the procedure for cleaning the Si substrate is outlined in Sect. 3.2.1. In Sect. 3.2.2, the preparation and characterization of Tl/Si(111)-(1×1) will be discussed. Furthermore, the symmetry of the surface is evaluated with regard to the spin degeneracies and nondegeneracies of the system and possibly arising spin textures.

3.2.1 Si(111)

Due to its applications in semiconductor electronics, Si(111) is a widely used substrate and standardized cleaning procedures can be found in literature, e.g., in Swartzentruber et al. (1989) and Senftleben et al. (2008). The Si(111) substrates used in this study were cut from a P-doped Si(111) wafer (n-type, $10^{15}/\mathrm{cm}^3$, 1–5 $\Omega\,\mathrm{cm}$), which has been chemically preoxidized. After insertion into the vacuum system, the sample and sample carrier were degassed at about 800 K for several hours, until the base pressure of the UHV chamber ($p \leq 1 \times 10^{-10}$ mbar) was restored. Next, the sample was heated at 1230 K for about 1 min by electron bombardment to remove the oxide layer. Subsequently, a flash to 1520 K for about 10 s was applied to remove the carbon contamination from the surface. Temperatures were measured with a pyrometer. A slow cooling down is necessary to allow a reordering of the surface atoms. Either of two cool-down procedures were applied: (A) Cooling down to room temperature at a constant rate of $5\,\mathrm{K\,s^{-1}}$. (B) Moderately fast cooling to 1170 K ($10\,\mathrm{K\,s^{-1}}$), slow cooling ($\leq 1\,\mathrm{K\,s^{-1}}$) through the (1×1)-(7×7) phase transition at about 1130 K (Höfer et al. 1995) to allow large 7×7 reconstructed domains to grow, and a more rapid cooling to room temperature. Both cool-down procedures led to clean Si(111)-(7×7) surfaces with high crystalline quality. The Si(111)-(7×7) reconstruction can be described within the dimer adatom stacking fault (DAS) model (Takayanagi et al. 1985), which is illustrated in Fig. 3.1a. In comparison with procedure A, method B produced even more homogenous surfaces. Frequent flashing led to a degradation of the crystalline quality. Therefore, the substrate has been replaced after a couple of preparations.

The cleanliness and crystalline quality were checked with LEED, AES and STM. The LEED pattern (see Fig. 3.1b) agrees well with the reciprocal surface lattice of the (7×7) reconstruction. The AES spectrum (see Fig. 3.1c) shows no traces of contaminants above the detection limit of the spectrometer. The STM image in Fig. 3.1d indicates terraces of about 50 nm size. STM scans of smaller areas in Figs. 3.1e, f reveal the occupied and the unoccupied adatom states of the (7×7) reconstruction, respectively (Hamers et al. 1986). The unit cell of the DAS model is in agreement with the unit cell of the observed structures in the STM images. The lighter- and darker-appearing triangular shaped areas in Fig. 3.1f are a result of the stacking fault in one half of the unit cell.

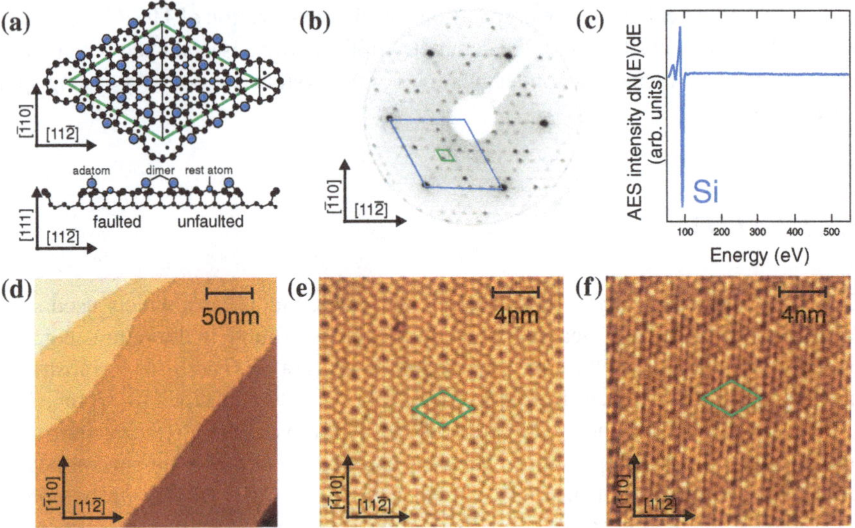

Fig. 3.1 **a** Structural model of the Si(111)-(7×7) surface according to the DAS model (Takayanagi et al. 1985). The unit cell is shown in *green* and can be divided into a faulted and an unfaulted half. **b** Inverted LEED image (primary electron beam energy $E_p = 54\,\text{eV}$) of the cleaned Si(111)-(7×7) surface. *Black* corresponds to high intensity. The (1×1) and the (7×7) unit cell of the reciprocal lattice are shown in *blue* and *green*, respectively. **c** Auger spectrum of the cleaned Si(111)-(7×7) surface for a primary energy of $E_p = 3\,\text{keV}$. The characteristic Si lines can be observed. No traces of contaminants are found. STM images of the cleaned Si(111)-(7×7) surface for **d** a large area ($200\times200\,\text{nm}^2$, $U = 1\,\text{V}$, $I = 0.1\,\text{nA}$) and smaller areas **e** ($20\times20\,\text{nm}^2$, $U = -1\,\text{V}$, $I = 0.1\,\text{nA}$) and **f** ($20\times20\,\text{nm}^2$, $U = 1\,\text{V}$, $I = 0.1\,\text{nA}$). Large terraces of about 50 nm size are observed. **e** and **f** reveal the empty and filled adatom states of the (7×7) reconstruction. For comparison, the (7×7) unit cell is drawn within the STM images

3.2.2 Tl/Si(111)-(1×1)

Preparation and Characterization

Tl/Si(111)-(1×1) films, schematically shown in Fig. 3.2a, were prepared by evaporating Tl from a Ta crucible onto the clean Si(111) substrate at a temperature of 570 K, similar to the recipe given in Sakamoto et al. (2006, 2009). During evaporation, the pressure was below 1×10^{-10} mbar. At a substrate temperature of 570 K, the formation of a 1 ML thick film is a self-terminating process. Excess Tl atoms do not stick to the surface.

The sharp diffraction pattern and low background intensity observed in the LEED image in Fig. 3.2b reflect a well ordered (1×1) structure. No traces of contaminants are detected with AES (see Fig. 3.2c). Figure 3.2d shows a STM image of the Tl/Si(111)-(1×1) sample. Large terraces (≈50 nm) imposed by the Si substrate are observed. Scans of smaller areas indicate only few defects similar to observations by Kocán et al. (2011). Other preparation methods as, e.g., presented in

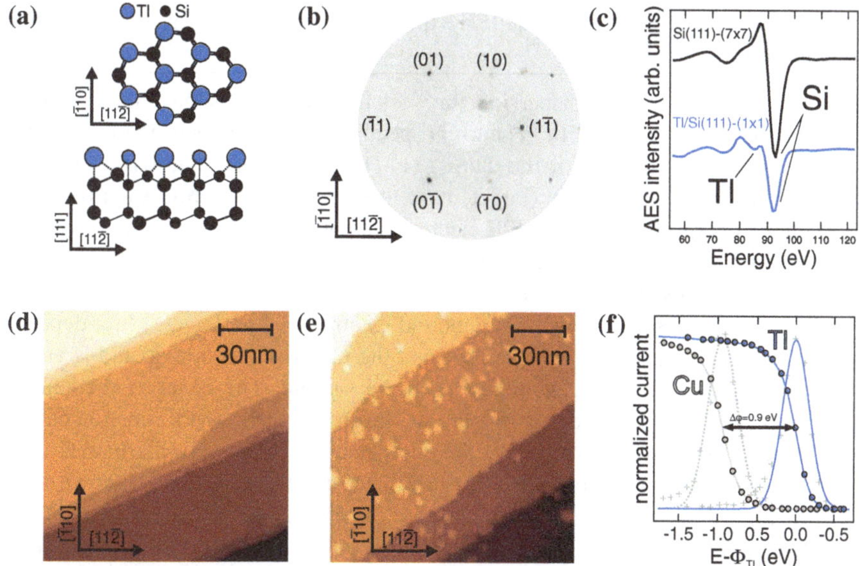

Fig. 3.2 **a** *Top view* and *side view* of the structural model of the Tl/Si(111)-(1×1) surface. **b** Inverted LEED image (primary electron beam energy $E_p = 85\,\text{eV}$) and **c** Auger spectrum ($E_p = 3\,\text{keV}$) of the clean Tl/Si(111)-(1×1) surface (*blue line*). The characteristic Tl lines coincide with the Si transitions. In comparison with the Auger spectrum of the clean Si substrate (*black line*), the adsorption of Tl alters the intensity of the Si-induced peaks and an additional Tl-induced peak appears. No contaminants are detected. **d** STM image of the Tl/Si(111)-(1×1) surface (150×150 nm^2, $U = -1\,\text{V}$, $I = 0.1\,\text{nA}$). Large terraces with a width of about 50 nm are observed. **e** STM image of the Tl/Si(111)-(1×1) surface with about 1.1 ML Tl (140×140 nm^2, $U = 1\,\text{V}$, $I = 0.1\,\text{nA}$). **f** Target-current-spectroscopy measurement (TCS) of the Tl/Si(111)-(1×1) surface (*blue markers* and *line*). The TCS measurement of the clean Cu(111) surface (*gray markers* and *line*) with $\Phi_{\text{Cu}} = 4.94\,\text{eV}$ (Takeuchi et al. 2001) serves as a reference. The difference of the onsets $\Delta\Phi = (0.90 \pm 0.05)\,\text{eV}$ is used to determine the work function $\Phi_{\text{Tl}} = (4.04 \pm 0.05)\,\text{eV}$ of the Tl/Si(111)-(1×1) surface

Kocán et al. (2011), were tested and led to similar film qualities. The homogeneity of the Si substrate directly influenced the homogeneity of the Tl/Si(111)-(1×1) films. Please note for future reference that Tl/Si(111)-(1×1) preparations on Si substrates cleaned with method A and B are referred to as preparation A and B, respectively.

Tl/Si(111)-(1×1) films with a small extra amount of Tl, e.g., 1.1 ML films, were grown by evaporating about two monolayer of Tl onto the Si substrate kept at room temperature. Subsequently, the sample was annealed to 500 K, while monitoring the diffraction pattern with the LEED experiment. With advancing annealing time the formation of a (1×1) structure was observed. Ceasing the annealing before a sharp unchanging diffraction pattern was observed led to Tl/Si(111)-(1×1) films with a small extra amount of Tl. The STM image in Fig. 3.2e shows a Tl/Si(111)-(1×1) film with a coverage of about 1.1 ML.

Additionally, with regard to SR-IPE experiments, two important aspects have to be considered.

(i) For mapping the $E(\mathbf{k}_{\parallel})$ dispersion, the work function of the sample is an essential quantity (see Sect. 2.1). Target current spectroscopy (TCS) is used to determine the work function Φ_{Tl} of the Tl/Si(111)-(1×1) samples. The basic principle is as follows. The sample (target) current is monitored, while electrons with defined energy from the SR-IPE electron source are guided onto the sample. A sample current is measured, if the kinetic energy E_{kin} of the electrons, i.e., the energy with respect to the vacuum level of the sample, is positive and there are electronic states, into which an unbound electron can propagate. E_{kin} depends on the potential between the sample and the GaAs photocathode and is varied by applying a retarding voltage U_R to the sample. The onset of the sample current is then used to determine the work function. The work function can be deduced, if the work function of the photocathode is known. Alternatively, the TCS spectrum can be compared to a TCS spectrum of a sample with known work function. Figure 3.2f shows the TCS spectra obtained for Tl/Si(111)-(1×1) and for Cu(111) [$\Phi_{\mathrm{Cu}} = 4.94$ (Takeuchi et al. 2001)]. With an onset difference of $\Delta\Phi = (0.90 \pm 0.05)$ eV, the work function of the Tl/Si(111)-(1×1) sample is determined to $\Phi_{\mathrm{Tl}} = (4.04 \pm 0.05)$ eV. This is in good agreement with previous studies, in which $\Phi_{\mathrm{Tl}} = 4$ eV is found (Hwang et al. 2007).

(ii) The second aspect concerns a possible voltage drop across the semiconducting sample, when applying current densities of 1–10 μA in SR-IPE or high photon fluxes in (S)ARPES (Himpsel 1990; Zhang et al. 2008). A voltage drop may be a result of a surface photovoltage (SPV) (Schroder 2001) or Schottky barriers (see, e.g., Tung 2014), which form at the back contact or the free surface. These affect the energy zero observed in the SR-IPE spectrum. Typically, the SR-IPE spectra are referenced to the Fermi level E_{F}. Usually, in SR-IPE, E_{F} is determined by measuring the Fermi level cut-off at a metallic surface with respect to the accelerating voltage U between the sample and the cathode (Dose et al. 1986). With a voltage drop U_D, the accelerating voltage becomes $U + U_D$ and E_{F} is found at a different position with respect to U. A correct energy calibration is, therefore, mandatory. A comparison of the Fermi level cut-off determined for the metallic Si(111)-(7×7) surface with the one found for a polycrystalline Ta sample shows no difference. Therefore, a voltage drop across the Si sample can be ruled out (Himpsel 1990). Furthermore, the SR-IPE spectra of Tl/Si(111)-(1×1) show no indication of a voltage drop.

Surface Symmetry

As discussed in Sect. 1.2, SOI may lift spin degeneracies of electronic states in crystalline solids. The lack of inversion symmetry at the surface in combination with strong potential gradients induced by heavy nuclei and asymmetric surface-state wavefunctions may give rise to spin-orbit-split surface states. Here, the $\mathbf{k}_{\parallel}$-dependent

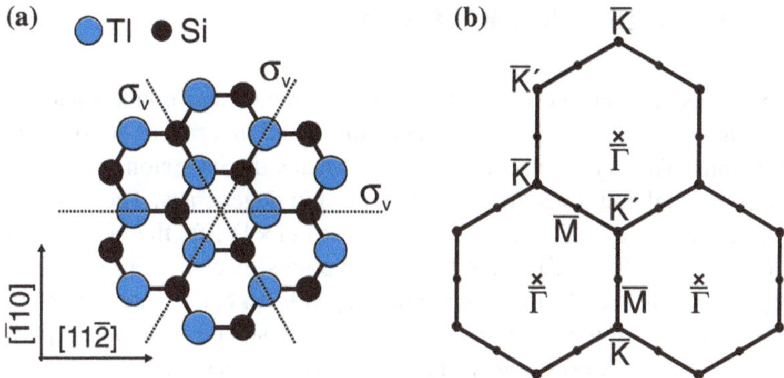

Fig. 3.3 **a** Structural model of the Tl/Si(111)-(1×1) surface. The mirror planes are indicated by *dashed lines*. **b** Illustration of the reciprocal lattice (*crosses*) of the Tl/Si(111)-(1×1) surface and three adjacent surface Brillouin zones with the relevant high-symmetry points (*circles*)

spin quantization axis is in a close relationship to the symmetry of the crystal. The symmetry of Tl/Si(111)-(1×1) is discussed with the help of Fig. 3.3, which shows the structural model of the Tl/Si(111)-(1×1) surface and the corresponding reciprocal lattice with three adjacent surface Brillouin zones. Taking the first Si layer into account, Tl/Si(111)-(1×1) can be viewed as honeycomb layered structure with threefold symmetry C_{3v} (Schönflies notation) belonging to the p3m1 space group (Hermann-Mauguin notation). The Tl/Si(111)-(1×1) surface possesses three mirror planes along the $\bar{\Gamma}\bar{\text{M}}$ directions, whereas along $\bar{\Gamma}\bar{\text{K}}$ ($\bar{\Gamma}\bar{\text{K}}'$) no mirror planes exist. Note that the basis of the surface lattice consists of two atoms, one Tl and one Si atom. This leads to a distinction between $\bar{\text{K}}$ and $\bar{\text{K}}'$. $\bar{\text{K}}$ and $\bar{\text{K}}'$ possess C_3 symmetry, while the $\bar{\text{M}}$ points are C_{1h} symmetric. Simple symmetry considerations are presented to provide a prospect of the possible spin textures of a surface state in compliance with the Tl/Si(111)-(1×1) surface and time-reversal symmetry.

A surface state of Tl/Si(111)-(1×1) has to be invariant under time reversal and the symmetry operations of the system, e.g., the mirror-plane operations (see Sect. 1.2). As a consequence, along $\bar{\Gamma}\bar{\text{M}}$, the spin quantization axis is fixed to the in-plane direction perpendicular to $\bar{\Gamma}\bar{\text{M}}$ (in-plane$_\perp$ spin-polarization direction). Along $\bar{\Gamma}\bar{\text{K}}$ ($\bar{\Gamma}\bar{\text{K}}'$) and $\bar{\text{M}}\bar{\text{K}}$ ($\bar{\text{M}}\bar{\text{K}}'$), in-plane$_\perp$ and out-of-plane spin polarization is allowed, whereas in-plane spin polarization parallel to $\mathbf{k}_\parallel$ is forbidden. Spin degeneracy is still preserved for the $\bar{\Gamma}$ and $\bar{\text{M}}$ points as these are time-reversal invariant momenta (see, e.g.,Dil et al. 2008). $\bar{\text{K}}$ and $\bar{\text{K}}'$ exhibit C_3 symmetry and are not time-reversal invariant momenta. This forces the spin-polarization direction to the out-of-plane spin-polarization direction (Oguchi and Shishidou 2009). For momenta in between $\bar{\Gamma}\bar{\text{M}}$, $\bar{\text{M}}\bar{\text{K}}$ ($\bar{\text{M}}\bar{\text{K}}'$) and $\bar{\Gamma}\bar{\text{K}}$ ($\bar{\Gamma}\bar{\text{K}}'$), the spin-polarization vector may, in principle, point in any direction.

As will be shown in the following sections, Tl/Si(111)-(1×1) features an unoccupied spin-orbit-split surface state, which can be traced throughout the complete surface Brillouin zone. Therefore, all predictions stated above can be tested.

3.3 Results for the $\bar{\Gamma}\bar{K}(\bar{\Gamma}\bar{K}')$ Direction

In the previous section, the possible spin degeneracies and nondegeneracies of a surface state of Tl/Si(111)-(1×1) have been discussed in terms of simple symmetry considerations. The crystal-vacuum interface breaks the inversion symmetry along the surface normal and, thus, lifts the in-plane spin degeneracy. The missing inversion symmetry in the surface plane of Tl/Si(111)-(1×1) lifts the out-of-plane spin degeneracy along $\bar{\Gamma}\bar{K}$ ($\bar{\Gamma}\bar{K}'$) and at the $\bar{K}$ ($\bar{K}'$) points. A closer investigation of the spin-dependent surface electronic structure depends on band structure calculations including spin-orbit interaction and, first and foremost, spin-resolved experiments. For the $\bar{\Gamma}\bar{K}$ ($\bar{\Gamma}\bar{K}'$) direction, both will be presented in this section.

3.3.1 Theoretical Predictions

We start with a discussion of the calculated Tl/Si(111)-(1×1) surface electronic structure. Several calculations of the spin-dependent surface electronic structure along the $\bar{\Gamma}\bar{K}$ direction of Tl/Si(111)-(1×1) can be found in literature (Lee et al. 2002; Ibañez-Azpiroz et al. 2011; Stolwijk et al. 2013). Here, band structure calculations are presented, which were performed by P. Krüger and are published in Stolwijk et al. (2013). The ground-state properties are obtained by using density-functional theory within the local-density approximation. A basis of Gaussian orbitals is employed together with pseudopotentials. These include scalar relativistic corrections and spin-orbit coupling (Stärk et al. 2011). The Tl/Si(111) surface is treated within a supercell approach using slabs with 18 Si substrate layers and a Tl adlayer. Relaxations of the topmost eight layers have been taken into account. Adsorption of Tl atoms in T_4 positions turns out to be energetically most favorable in agreement with former calculations (Lee et al. 2002). For determining the quasiparticle band structure, the self-energy operator is constructed within the GW approximation (Hybertsen and Louie 1988; Sommer et al. 2012).

Figure 3.4 presents the band structure calculations along $\bar{\Gamma}\bar{K}\bar{M}$. Open circles illustrate the in-plane spin polarization perpendicular to $\mathbf{k}_{\parallel}$ (in-plane$_{\perp}$ spin polarization) (left panel) and the out-of-plane components of the spin-polarization vector (right panel) as defined in the insets. No in-plane spin polarization parallel to $\mathbf{k}_{\parallel}$ occurs (in-plane$_{\parallel}$ spin polarization). The diameter of the circles corresponds to the magnitude of the respective polarization. The gray-shaded area denotes the projected bulk bands. In accordance with band structure calculations in Ibañez-Azpiroz et al. (2011), one occupied (S_1 and S_2) and one unoccupied (S_3 and S_4) spin-orbit-split surface state is found within the projected bulk-band gap of the silicon substrate. The occupied surface state primarily stems from the dangling bonds of the surface Si atoms, whereas the unoccupied surface state originates from the Tl p orbitals (see also discussion in Sect. 3.3.4). As discussed in Sect. 3.2, the 2D symmetry of the hexagonal system, belonging to the p3m1 space group, leads to $\bar{K}$ and $\bar{K}'$ points with C_3 symmetry. This forces the spin-polarization vector to the out-of-plane direction (Oguchi and Shishidou 2009; Sakamoto et al. 2009; Ibañez-Azpiroz et al. 2011). Ultimately, the spin-polarization vectors of all four surface-state components rotate from the in-plane$_\perp$ direction around $\bar{\Gamma}$ to the out-of-plane direction at $\bar{K}(\bar{K}')$.

3.3.2 *Occupied Surface States*

So far, measurements on the surface electronic structure of Tl/Si(111)-(1×1) have been focused on the occupied part (Lee et al. 2002; Sakamoto et al. 2009, 2013). Figures 3.5a, b present high-resolution ARPES and SARPES data reproduced from Sakamoto et al. (2009, 2013), respectively. In Fig. 3.5b, the SARPES data are presented along with the projected bulk bands (gray-shaded area) and the calculated surface-state bands S_1 and S_2 from the calculations in Fig. 3.4.

The (S)ARPES data uncover the occupied surface-state components S_1 and S_2. Around $\bar{\Gamma}$, S_1 and S_2 are in-plane$_\perp$ spin polarized. At about four fifths of the $\bar{\Gamma}\bar{K}$ direction, the spin-polarization direction rotates to the direction perpendicular to the surface. At $\bar{K}$, S_1 and S_2 possess almost complete out-of-plane spin polarization and are split in energy by about 0.25 eV.

Results for Tl/Si(111)-(1×1) surfaces prepared in this work and obtained with the ARPES setup described in Sect. 2.2 are presented in Fig. 3.6a, b, which show spectra (open circles) taken at $\bar{\Gamma}$ and around $\bar{K}$, respectively. Note that the Si substrates used in Sakamoto et al. (2009) and in this work were cut from the same or equivalent wafers[1] and that the ARPES experiments have been conducted for the same quantum energy of $\hbar\omega = 21.22$ eV. In comparison with the data of Sakamoto et al. (2009)

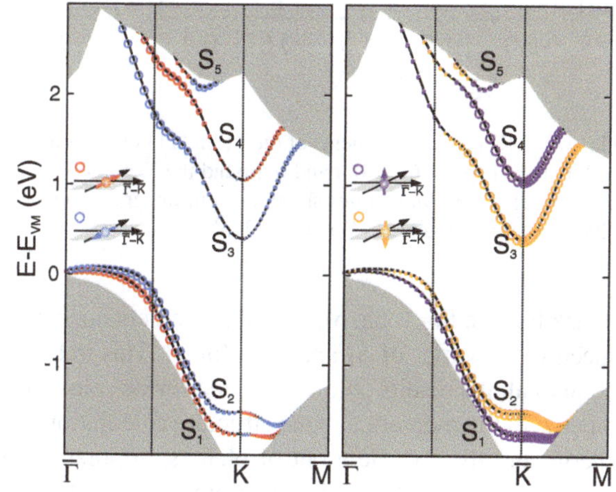

Fig. 3.4 Quasiparticle band structures including spin-orbit coupling illustrating the in-plane$_\perp$ (*left hand side*) and out-of-plane (*right hand side*) polarization components. The diameter of the *circles* is proportional to the spin polarization with a maximum degree of 100 %, e.g., for S_4 at $\bar{K}$. The energy scale refers to the valence-band maximum E_{VM}. The *gray-shaded area* illustrates the projected bulk bands

[1]The wafers were provided by K. Sakamoto.

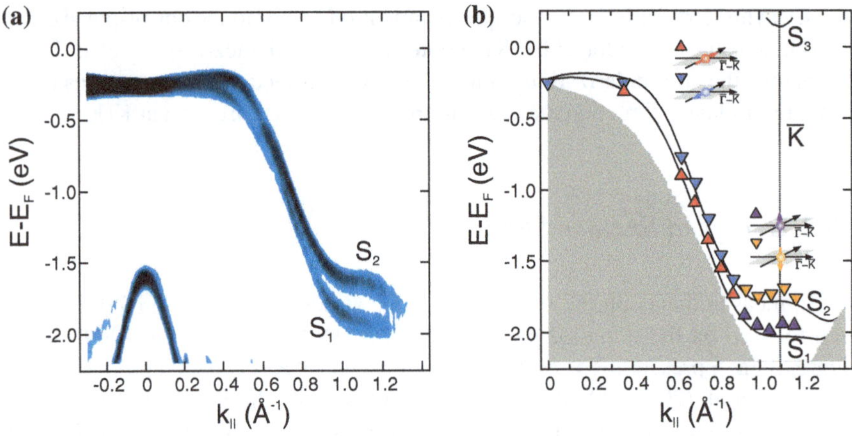

Fig. 3.5 **a** ARPES data published in Sakamoto et al. (2013) and **b** SARPES data published in Sakamoto et al. (2009) showing the spin-split occupied surface-state bands S_1 and S_2. The data has been reused with permission from K.S. Sakamoto. The spin-polarization vectors of S_1 and S_2 rotate from the in-plane$_\perp$ direction around $\bar{\Gamma}$ to the out-of-plane spin-polarization direction at $\bar{K}$

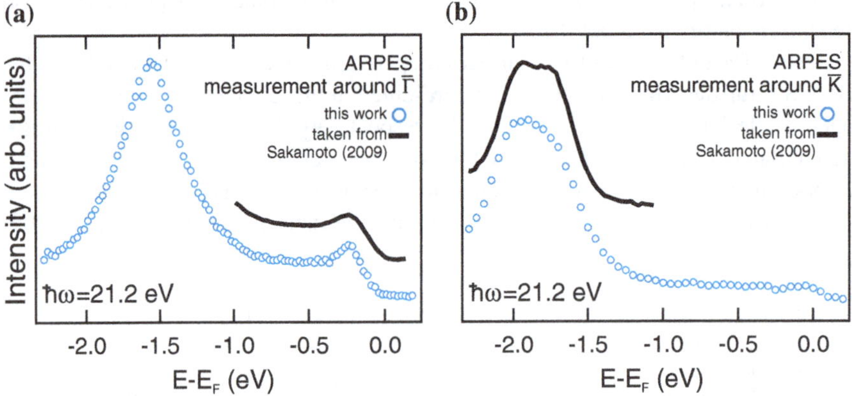

Fig. 3.6 ARPES measurements of the Tl/Si(111)-(1×1) sample prepared in this work (*blue markers*) **a** at the $\bar{\Gamma}$ point ($\theta = 0°$) and **b** around the $\bar{K}$ point ($\theta \approx 32°$) obtained with a photon energy of $\hbar\omega = 21.22\,\text{eV}$. *Black lines* show the spin-integrated photoemission results of Sakamoto et al. (2009) at $\bar{\Gamma}$ and around $\bar{K}$ ($\hbar\omega = 21.22\,\text{eV}$)

(black lines in Fig. 3.6a, b), a good agreement concerning the energetic positions and spectral intensities of S_1 and S_2 is found. This indicates similar surface qualities.[2]

In total, the results on the occupied surface electronic structure are in good agreement with the theoretical band structure calculations. However, the large splitting in energy and the complete out-of-plane spin polarization occur far below the Fermi level. The occupied state is, therefore, hardly relevant for spin-dependent transport

[2]Differences may be a result of the different energy resolutions of the experiments.

phenomena. On the other hand, for the unoccupied electronic structure, theory predicts a surface state with rotating spin-polarization vector and exceptionally large splitting in the vicinity of E_F (see Fig. 3.4). The spin- and angle-resolved inverse-photoemission experiments presented in this thesis put these predictions to the test.

3.3.3 Unoccupied Surface States

The first aim of the experimental study on the unoccupied surface electronic structure of Tl/Si(111)-(1×1) is the identification of the theoretically predicted unoccupied surface states along $\bar{\Gamma}\bar{K}$ (see Fig. 3.4). This is achieved by means of SR-IPE experiments with the ROSE, with the spin sensitivity set either to the out-of-plane or in-plane$_\perp$ spin-polarization direction. Unless otherwise noted, spectra have been normalized to 100 % spin polarization of the electron beam.

Figure 3.7a presents SR-IPE spectra with sensitivity to the in-plane$_\perp$ spin-polarization direction taken at various angles of electron incidence θ along $\bar{\Gamma}\bar{K}$. Red up- and blue down-pointing triangles represent the in-plane$_\perp$ spin-polarization direction. In Fig. 3.7b, SR-IPE spectra with sensitivity to the out-of-plane spin-polarization direction are shown. Purple up- and orange down-pointing triangles denote states with out-of-plane spin polarization parallel and antiparallel to the surface normal, respectively. Note that experimental information on the out-of-plane spin polarization is only accessible for angles θ unequal zero, and the effective spin polarization increases with higher θ. Therefore, the spectra at $\theta = 0°$ and 5° have not been normalized to a 100 % polarized electron beam. Importantly, for corresponding angles, the spin-integrated spectra of the in-plane$_\perp$ and out-of-plane sensitive experiments are almost identical (see comparison of spin-integrated spectra in Fig. 3.7c).This verifies the same sample condition and position in k space and highlights the beauty of the ROSE.[3] The presented data are obtained from Tl/Si(111)-(1×1) samples prepared with method A (see Sect. 3.2.1).

To give an overview of the detected states, spectral features observed in the SR-IPE spectra are translated into $E(\mathbf{k}_\parallel)$ plots in Fig. 3.8a, e for the in-plane$_\perp$ and out-of-plane spin channels, respectively. The energetic positions of the spectral features are derived by a peak analysis routine as described in Sect. 2.1. Note that the $E(\mathbf{k}_\parallel)$ plots contain additional data points, which are obtained from additional spectra presented in Appendix A.1 and spectra from samples, which are obtained from a different preparation method (preparation B). Furthermore, Fig. 3.8b, c, f, g present $E(\mathbf{k}_\parallel)$ false color plots of the second derivatives of the SR-IPE spectra. These are obtained as described in Appendix A.2. For comparison, the band structure calculations from Fig. 3.4 are shown in Fig. 3.8d, h with respect to the Fermi level and with the respective spin-polarization components. E_F lies about 0.25 eV above the highest occupied state S_2 at $\bar{\Gamma}$, as measured by ARPES (see Sect. 3.3.1), and about 0.2 eV below the lowest

[3]For $\theta = 50°$, deviations are observed, which may be a result of slightly different angles of electron incidence.

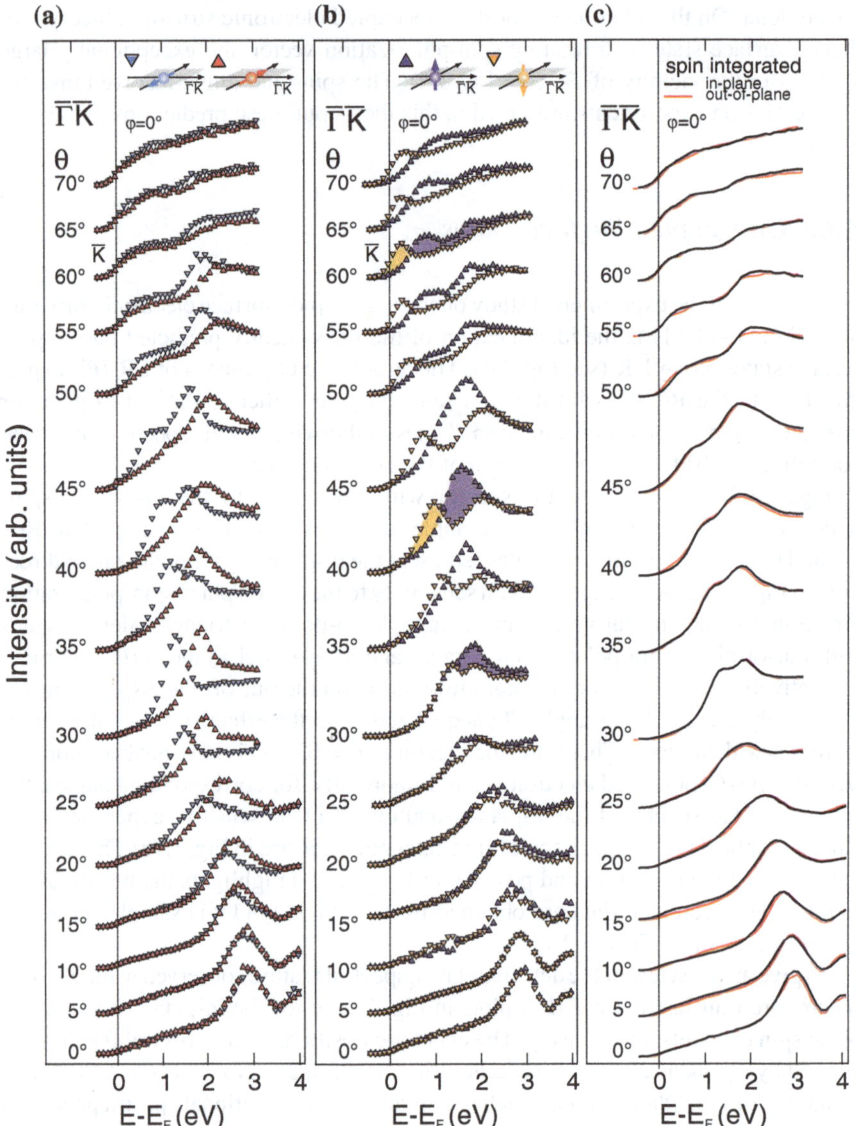

Fig. 3.7 SR-IPE spectra of Tl/Si(111)-(1×1) (preparation A) along $\bar{\Gamma}\bar{K}$ sensitive to the **a** in-plane$_{\perp}$ (*red* up- and *blue* down-pointing *triangles*) and **b** out-of-plane spin-polarization direction (*purple* up- and *orange* down-pointing *triangles*). **c** *Red* and *black lines* represent the spin-integrated spectra from (**a**) and (**b**), respectively. θ and φ denote the angle of electron incidence and the sample azimuth, respectively

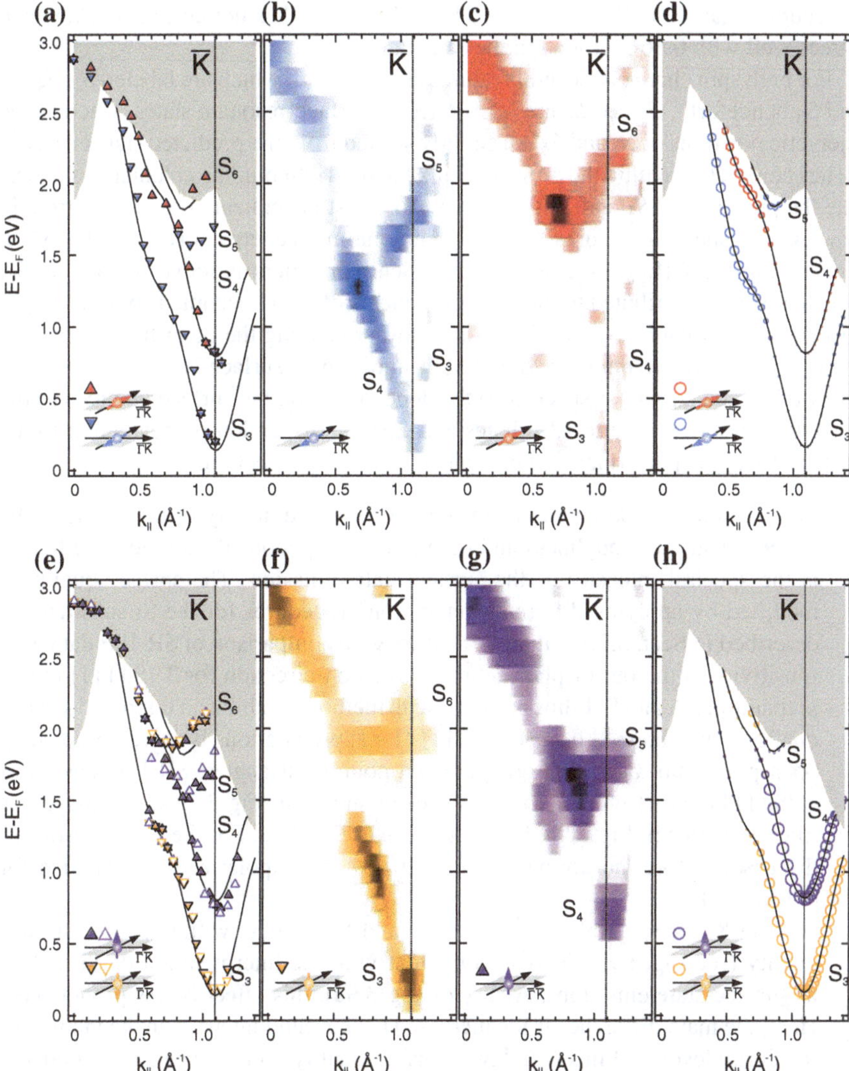

Fig. 3.8 **a, e** $E(\mathbf{k}_\parallel)$ plot derived from spectra taken along $\bar{\Gamma}\bar{K}$ with sensitivity to the in-plane$_\perp$ (*red* up- and *blue* down-pointing *triangles*) and out-of-plane spin-polarization direction (*purple* up- and *orange* down-pointing *triangles*), respectively. *Filled* and *open triangles* are obtained for Tl/Si(111)-(1×1) preparations A and B, respectively. *Solid lines* represent the calculated surface-state bands, and the *gray-shaded area* the projected bulk bands from the band structure calculation shown in (**d**) and (**h**) (see also Fig. 3.4). **b, c, f, g** *False color plots* of the second derivatives of the SR-IPE spectra for the respective spin channels

unoccupied state S_3 at $\bar{\text{K}}$. ΔE between S_2 at $\bar{\Gamma}$ and S_3 at $\bar{\text{K}}$ amounts to 0.45 eV, which agrees well with 0.36 eV found in the calculations.

For both spin channels, several features are observed, which are labeled S_3, S_4, S_5 and S_6 henceforth. S_5 and S_6 are interpreted as surface-resonant states, whereas the energetic positions of S_5 and S_6 are slightly shifted from the predicted state energies. S_6 lies completely within the projected bulk bands. Tight binding calculations underline the presence of S_5 and S_6 (see Fig. 3.12). Most prominent are the two strongly downward dispersing features S_3 and S_4 with minimum energy at $\bar{\text{K}}$ ($k_{\parallel} \approx 1.1\,\text{Å}^{-1}$). Concerning the $E(\mathbf{k}_{\parallel})$ dispersion, an excellent agreement is found for S_3 and S_4, which are easily attributed to the two components of the unoccupied spin-orbit-split surface state. Notably, S_3 and S_4 are also observed along the other high-symmetry directions. In order to pinpoint S_3 and S_4 as unoccupied surface-state components, a study of S_3 and S_4 with respect to (i) the dependence on the surface quality, (ii) the behavior upon exposure to adsorbates and (iii) the dispersion perpendicular to the surface $E(\mathbf{k}_{\perp})$ is performed and presented for the $\bar{\Gamma}\bar{\text{K}}$ direction.

(i) Surface states are known to react sensitively to the quality of the surface: The lower the surface roughness and the more homogenous the surface, the higher is the spectral intensity of the surface-state emission. The surface quality is modified by applying different preparation procedures for the Si substrate, as described in Sect. 3.2.1. Figure 3.9a–c shows a comparison of SR-IPE data with sensitivity to the out-of-plane spin-polarization direction for Tl/Si(111)-(1×1) preparations A and B. Films prepared with method B exhibit an improved surface quality. Data obtained from Tl/Si(111)-(1×1) preparations B are represented by open purple up- and open orange down-pointing triangles. SR-IPE spectra of Tl/Si(111)-(1×1) preparations A (already shown in Fig. 3.7) are displayed as filled triangles in Fig. 3.9a, b, and as lines in Fig. 3.9c. For the same reasons as discussed before, no normalization to 100 % beam polarization is applied for $\theta = 0°$ and 5°.
Around $\bar{\text{K}}$, the spectral intensities of S_3 and S_4 increase with improved surface quality (see Fig. 3.9b). This is consistent with the interpretation of S_3 and S_4 as surface-state emissions. At $\bar{\Gamma}$ (see Fig. 3.9a), this effect is less pronounced. Here, the main difference is the increased background intensity in the vicinity of the Fermi level for films with lower surface quality. This can be understood as a consequence of an enhanced inelastic scattering due to the lower surface quality. For spectra taken along $\bar{\Gamma}\bar{\text{K}}$ (Fig. 3.9c), the same statements can be made. Note that the surface quality only affects the spectral intensities. The $E(\mathbf{k}_{\parallel})$ dispersion is almost identical (compare open with filled triangles in Fig. 3.8e).

(ii) A typical approach to identify surface states is to expose the surface to adsorbates and, thereby, intentionally increase the amount of impurities. Typically, an exposure to a few Langmuir L (1L= 1.33×10^{-4} Pa s) of, e.g., oxygen (O_2) or hydrogen (H_2), already suffices to completely quench surface-state emissions (see for example Altmann et al. 1985; Reihl et al. 1984). Figure 3.9d shows SR-IPE measurements sensitive to the out-of-plane spin-polarization direction around $\bar{\text{K}}$, which are obtained after an exposure to 100 L H_2, to 500 L O_2 and to

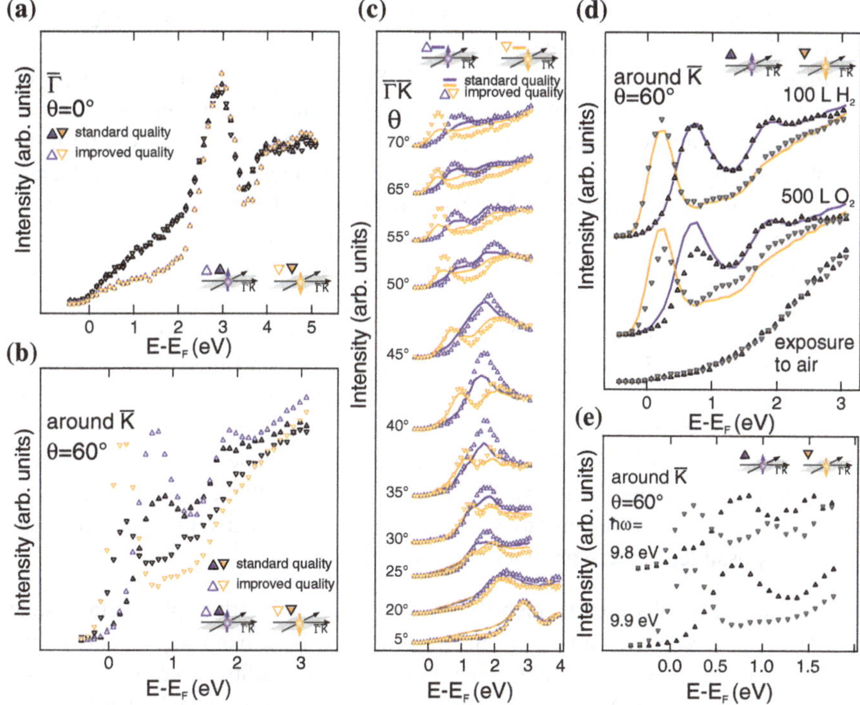

Fig. 3.9 SR-IPE spectra with sensitivity to the out-of-plane spin-polarization direction **a** at $\bar{\Gamma}$, **b** around $\bar{K}$ and **c** along $\bar{\Gamma}\bar{K}$, obtained from Tl/Si(111)-(1×1)-A (*filled purple* up- and *orange* down-pointing *triangles* in (**a**) and (**b**), *lines* in (**c**)) and Tl/Si(111)-(1×1)-B (*open purple* up- and *orange* down-pointing *triangles*). **d** SR-IPE spectra around $\bar{K}$ (preparation B) for the clean (*solid lines*) and adsorbate-covered surfaces (*filled purple* up- and *orange* down-pointing *triangles*). **e** SR-IPE spectra around $\bar{K}$ (preparation B) for two mean detection energies $\hbar\omega = 9.9\,\text{eV}$ (*lower* spectra) and $\hbar\omega = 9.8\,\text{eV}$ (*upper* spectra)

air. The starting point of each adsorbate experiment is the clean Tl/Si(111)-(1×1) (preparation B, shown as lines).

Exposure to 100 L H_2 and 500 L O_2 has almost no affect on the observed spectral intensities of S_3 and S_4. At the first glance, this observation is in contradiction with the interpretation of S_3 and S_4 as surface-state emissions. However, in the case of Tl/Si(111)-(1×1), the robust behavior of the surface states to foreign adsorbates reflects the electronically saturated nature of the surface dangling bonds (Lee et al. 2002). The dangling bonds of the Si atoms are saturated by the electrons from the Tl adatoms and the surface becomes extremely inert. No state emissions are observed after exposure to air.

(iii) Surface states are two-dimensional states and, thus, possess no dispersion perpendicular to the surface (Lüth 2001). Hence, to identify surface-state emissions in SR-IPE experiments, one possibility is to alter the momentum perpendicular to the surface $\mathbf{k}_\perp$. In the isochromat mode of the SR-IPE experiment, this can be realized by the use of different photon detection energies, in analogy to the use of different excitation energies in (S)ARPES setups.

Figure 3.9e presents measurements of Tl/Si(111)-(1×1) (preparation B) with sensitivity to the out-of-plane spin-polarization direction around $\bar{\mathrm{K}}$ with two different mean detection energies, $\hbar\omega = 9.9\,\mathrm{eV}$ and $\hbar\omega = 9.8\,\mathrm{eV}$. These are obtained by varying the temperature of the entrance window of the Geiger counter as described in Sect. 2.2. Note that counter $C1$ was used for the measurement with $\hbar\omega = 9.9\,\mathrm{eV}$, whereas the spectrum with $\hbar\omega = 9.8\,\mathrm{eV}$ was taken with a different counter ($C2$). As $C1$ and $C2$ detect the photons under different angles, the different spectral intensities are likely a result of the anisotropic angular distribution of the surface-state emission. Also, the different energy resolutions, $\Delta E = 350\,\mathrm{meV}$ for $\hbar\omega = 9.9\,\mathrm{eV}$ and $\Delta E = 250\,\mathrm{meV}$ for $\hbar\omega = 9.8\,\mathrm{eV}$ (see Sect. 2.3.2), can explain differences. A change of the detection energy by $0.1\,\mathrm{eV}$ may strongly alter the energetic positions of bulk-state emissions (see, e.g., Wissing et al. 2013). Irrespective of the detection energy, S_3 and S_4 appear at the same energetic positions. This is in agreement with the interpretation of S_3 and S_4 as surface-state emissions. A direct comparison with bulk-state emissions is not possible, as these are not observed.

In total, the results clearly identify S_3 and S_4 as spin-orbit-split surface state of Tl/Si(111)-(1×1). S_3 and S_4 lie in the projected bulk band gap, are sensitive to the surface quality and do not depend on $\mathbf{k}_\perp$. The results agree well with the theoretical calculations. Moreover, adsorbate experiments reveal the high robustness of the unoccupied surface state.

3.3.4 Spin Texture

Quasiparticle band structure calculations predict a rotation of the spin-polarization vector of the unoccupied surface-state components from the in-plane$_\perp$ to the out-of-plane spin-polarization direction. In the following, the (i) in-plane$_\perp$ and (ii) out-of-plane spin-polarization components of S_3 and S_4 are addressed with the help of the SR-IPE measurements and the $E(\mathbf{k}_\parallel)$ plots presented in Figs. 3.7 and 3.8.

(i) First, the in-plane$_\perp$ spin polarization is evaluated. At $\bar{\Gamma}$ ($\theta = 0°$), the two spin states are degenerate. Along $\bar{\Gamma}\bar{\mathrm{K}}$, i.e., with increasing angles up to $\theta = 50°$, S_3 and S_4 become spin split and exhibit in-plane$_\perp$ spin polarization. Around $\bar{\mathrm{K}}$ ($\theta = 60°$–$65°$), the in-plane$_\perp$ spin polarization vanishes, as demanded by symmetry considerations. In this region, due to the intrinsic and experimental energy broadening, a splitting of the states is not resolved.

To test the $\mathbf{k}_\parallel$ dependence of the in-plane$_\perp$ spin polarization, SR-IPE spectra are taken for positive and negative angles of electron incidence ($\theta = 30°$ and $\theta = -30°$) along $\bar{\Gamma}\bar{\mathrm{K}}$ and $\bar{\Gamma}\bar{\mathrm{K}}'$. Measurements along $\bar{\Gamma}\bar{\mathrm{K}}$ and $\bar{\Gamma}\bar{\mathrm{K}}'$ are realized by rotating the azimuth φ of the sample by $60°$. For spectra taken at $\theta = 30°$(compare upper spectra in Fig. 3.10a, b), S_3 and S_4 appear at the

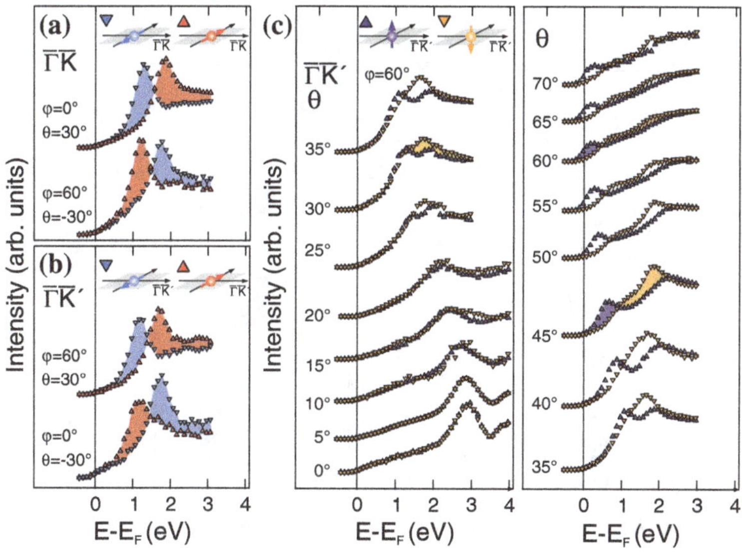

Fig. 3.10 SR-IPE spectra of Tl/Si(111)-(1×1) (preparation A) along (**a**) $\bar{\Gamma}\bar{K}$ and (**b**), (**c**) $\bar{\Gamma}\bar{K}'$ sensitive to the (**a**), (**b**) in-plane$_\perp$ (*red* up- and *blue* down-pointing *triangles*) and (**c**) out-of-plane spin-polarization direction (*purple* up- and *orange* down-pointing *triangles*)

same energy with the same spin-polarization direction, irrespective of the high-symmetry direction. Measurements at negative angles of electron incidence (compare lower spectra in Fig. 3.10a, b) detect S_3 and S_4 with reversed spin polarization. The behavior of the in-plane$_\perp$ spin polarization is in accordance with time-reversal symmetry and the Rashba-Bychkov model.

(ii) Along $\bar{\Gamma}\bar{K}$, S_3 and S_4 exhibit out-of-plane spin polarization for $\theta \geq 25°$. Figure 3.10c presents SR-IPE measurements along $\bar{\Gamma}\bar{K}'$ ($\varphi = 60°$). Compared to the measurements along $\bar{\Gamma}\bar{K}$, the same $E(\mathbf{k}_\parallel)$ dispersion is observed (compare data in Figs. 3.10d and 3.8). Notably, the features appear with opposite spin polarization (compare spectra in Figs. 3.7b and 3.10c). This is in accordance with time-reversal symmetry and the C_{3v} symmetry of the surface. It should be noted that measurements along $\bar{\Gamma}\bar{K}$ and $\bar{\Gamma}\bar{K}'$ are not equivalent, in that a rotation of 60° is not a symmetry operation of the system (see Fig. 3.3). Differences in the spectral intensities of the observed features are, thus, interpreted as a consequence of the threefold symmetry of the surface.

An overall picture of the spin polarization of the unoccupied surface-state components is found. Around $\bar{\Gamma}$, S_3 and S_4 exhibit only in-plane$_\perp$ spin polarization. This in-plane$_\perp$ spin polarization is retained up to $k_\parallel \approx 0.85\,\text{Å}^{-1}$ ($\theta = 45°$). For higher $k_\parallel$ values, it decreases and ultimately vanishes at $\bar{K}$. Out-of-plane spin polarization first appears at $k_\parallel \approx 0.6\,\text{Å}^{-1}$ ($\theta = 25°$). At the $\bar{K}$ point, S_3 and S_4 are almost completely out-of-plane spin polarized. In total, a rotation of the spin-polarization vector is observed. This coincides with the theoretical findings. As a detail it is noted that

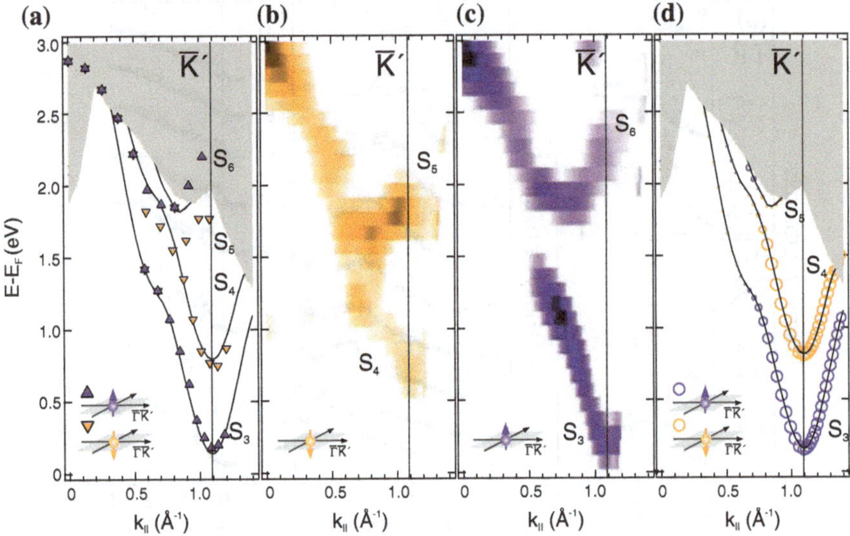

Fig. 3.11 **a** $E(\mathbf{k}_{\parallel})$ plot derived from spectra taken along $\bar{\Gamma}\bar{\mathrm{K}}'$ with sensitivity to the out-of-plane spin-polarization direction (*purple* up- and *orange* down-pointing *triangles*). *Solid lines* show the calculated surface-state bands, and the *gray-shaded area* represent the projected bulk bands from the band structure calculation shown in (**d**). **b**, **d** *False color plots* of the second derivatives of the SR-IPE spectra

along $\bar{\Gamma}\bar{\mathrm{K}}$, the out-of-plane spin-polarization direction is, first, parallel to the surface normal (see S_3 for $\theta = 25°$) and reverses for higher $k_{\parallel}$ values (see S_3 for $\theta \geq 35°$). The theoretical calculations predict a similar behavior.

In principle, the rotation of the spin-polarization vector can already be understood by applying a simple tight binding model, taking into account the Si dangling bond and the Tl p_x, p_y, p_z orbitals. Figure 3.12 presents the tight binding calculations (a) without and (b), (c) with interaction of the Tl adlayer with the Si substrate and spin-orbit interaction.[4] In Fig. 3.12b, c, the in-plane$_{\perp}$ and out-of-plane spin-polarization components are drawn as circles on top of the respective bands, respectively. The circle diameter denotes the magnitude of the spin polarization.

Without interactions, four bands can be identified, which originate from the Si dangling bond D_{Si} (blue dashed line in Fig. 3.12a), the Tl p_x and p_y orbitals T_1 and T_2 (black lines in Fig. 3.12a), and the Tl p_z orbital T_3 (red line in Fig. 3.12a). Spin-orbit interaction and interaction of the Tl adlayer with the substrate lead to a hybridization and a spin splitting of the bands. S_1 and S_2 primarily stem from the Si dangling bond, whereas S_3 to S_8 stem from the Tl orbitals. S_3 and S_4 correspond to the unoccupied surface state, S_5 and S_6 to the surface-resonant state found in the experiment. Interaction of the Tl p_z orbital with the Si substrate induces a potential gradient along the surface normal, resulting in an in-plane$_{\perp}$ spin polarization.

[4]The tight binding calculations were performed by Krüger (2013d).

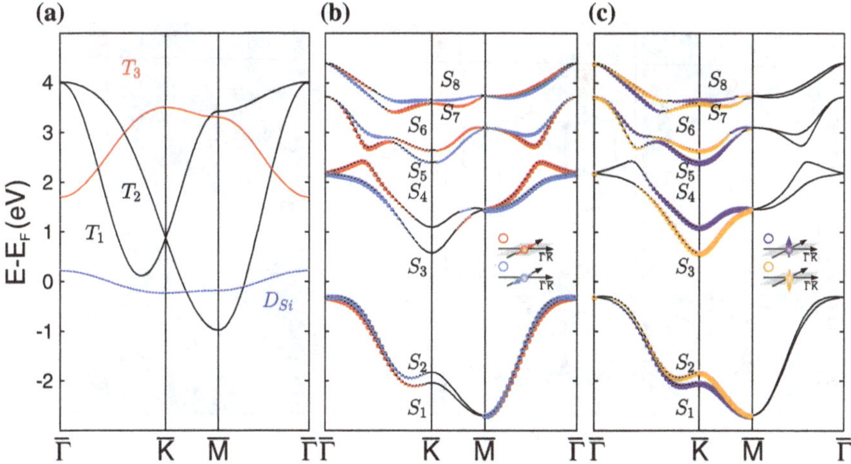

Fig. 3.12 Tight binding calculations of the Tl/Si(111)-(1×1) surface taking into account the Si dangling bond D_{Si} (*blue dashed line* in **a**), the Tl p_x and p_y orbitals T_1 and T_2 (*black lines* in **a**) and the Tl p_z orbital T_3 (*red line* in **a**). In (**a**), adlayer-substrate interaction and spin-orbit interaction are not included within the calculation. With adlayer-substrate interaction and spin-orbit interaction for the (**b**) in-plane$_\perp$ (*red* and *blue circles*) and (**c**) out-of-plane spin-polarization direction (*purple* and *orange circles*). The diameter of the *circles* is proportional to the spin polarization with a maximum degree of 100 %, e.g., for S_4 at $\bar{K}$

Interaction of the p_x, p_y orbitals with the substrate induces an in-plane inversion asymmetry resulting in an out-of-plane spin polarization. As the surface states S_3 and S_4 are predominantly of p_z character around $\bar{\Gamma}$ and purely of p_x, p_y character at $\bar{K}$, a transition from in-plane to out-of-plane spin polarization takes place. At $\bar{K}$, these states are completely spin polarized, since spin-orbit coupling does not intermix spin-up with spin-down components of p_x and p_y orbitals. Charge distribution plots obtained from calculations within the local-density approximation[5] of the unoccupied surface-state component S_3 at two points in k space comply with this model (see Fig. 3.13a). Halfway between $\bar{\Gamma}$ and $\bar{K}$ (marked by vertical lines in Fig. 3.4), the charge distribution is predominately asymmetric along the surface normal, which is related to an in-plane spin polarization. On the other hand, at $\bar{K}$, in-plane asymmetry is dominant, which relates to an out-of-plane spin polarization.

3.3.5 Conclusion

In conclusion, the experimental study of the unoccupied electronic structure of Tl/Si(111)-(1×1) along the $\bar{\Gamma}\bar{K}$ ($\bar{\Gamma}\bar{K}'$) direction reveals an unoccupied spin-orbit-

[5] Krüger carried out the calculations (Krüger 2013b), which are partly published in Stolwijk et al. (2013).

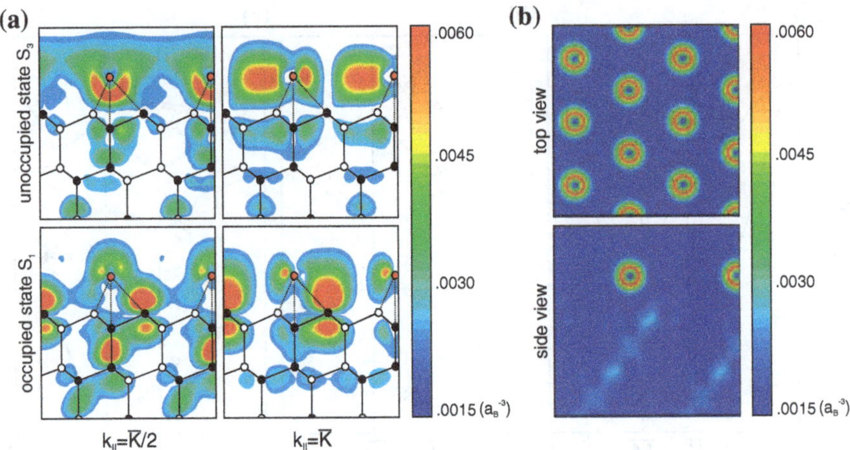

Fig. 3.13 **a** Charge distribution plots of the occupied and unoccupied surface-state component S_1 and S_3 at two points in k space $\bar{K}/2$ and $\bar{K}$. **b** *Top view* (*upper figure*) and *side view* (*lower figure*) of the k-integrated valence-band charge distribution of Tl/Si(111)-(1×1)

split surface state with unique properties, which are in excellent agreement with theory. It is unambiguously shown that the spin-polarization vectors of the unoccupied surface-state components rotate from the in-plane$_{\perp}$ spin-polarization direction to the direction perpendicular to the surface. As will be discussed in the following section, this leads to spin-split valleys with almost complete spin polarization at the $\bar{K}$ ($\bar{K}'$) points.

3.4 Results Around $\bar{K}$

In the following, the evaluation of the unoccupied spin-orbit-split surface state focuses on the electronic structure around $\bar{K}$ ($\bar{K}'$). Figure 3.14a–c and d–f shows the obtained data in the vicinity of $\bar{K}$ and $\bar{K}'$, respectively.

3.4.1 *Unoccupied Valleys*

Around $\bar{K}$ and $\bar{K}'$, the surface-state components S_3 and S_4 approach E_F and form spin-split pockets. In Fig. 3.15a, d, the dispersion around $\bar{K}$ is further analyzed by spin-resolved constant-energy maps at $E \approx 0.75$ eV and $E \approx 0.35$ eV (see dashed lines in Fig. 3.14), respectively. For each point of the constant-energy plots, θ and φ are set, so that the SR-IPE measurement (out-of-plane sensitivity) is carried out at a distinct value for k_x and k_y. The corresponding spin-asymmetry maps are presented

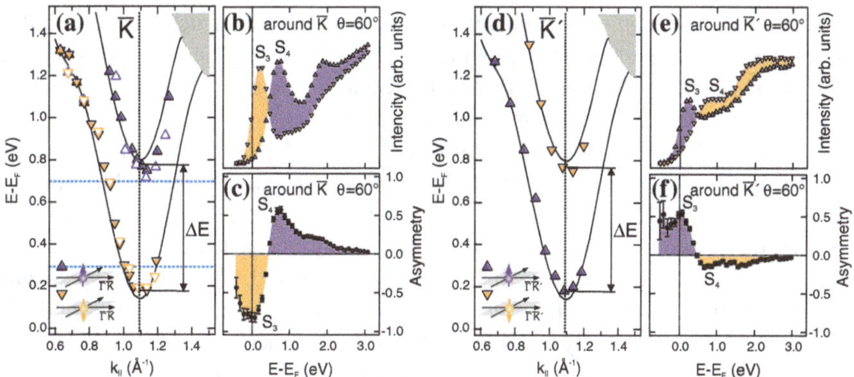

Fig. 3.14 Close-up of the surface electronic structure of Tl/Si(111)-(1×1) around **a** $\bar{K}$ and **d** $\bar{K}'$ according to Figs. 3.8 and 3.11. *Blue dashed lines* denote the cuts of the constant-energy maps in Fig. 3.15. **b** and **e** shows SR-IPE spectra with sensitivity to the out-of-plane spin polarization around $\bar{K}$ and $\bar{K}'$, respectively, with corresponding out-of-plane spin asymmetry shown in (**c**) and (**f**)

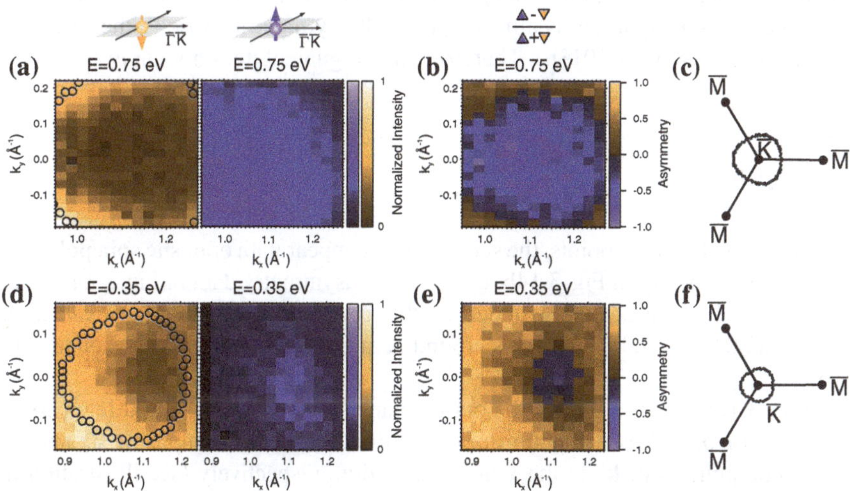

Fig. 3.15 Constant-energy maps of Tl/Si(111)-(1×1) (preparation B) around $\bar{K}$ at **a** $E \approx 0.75$ eV and **d** $E \approx 0.35$ eV. The *open circles* show the calculated dispersion of the lower lying surface-state component S_3 (see **c** and **f**). **b** and **e** shows the corresponding spin-asymmetry data. Calculated constant-energy maps at **c** $E \approx 0.75$ eV and **f** $E \approx 0.35$ eV (Krüger 2013a)

in Fig. 3.15b, e. At $E \approx 0.75$ eV, the constant-energy scan crosses the surface-state component S_3 and grazes the bottom of S_4. At $E \approx 0.35$ eV, only the lower lying surface-state component S_3 is measured. The results reveal S_3 as a ring-like feature, which is in good agreement with theoretical calculations (see Fig. 3.15f) (Krüger 2013a). The trigonal warping is found to be a consequence of the threefold symmetry. A trigonal warping is also predicted for MoS_2 (Kormányos et al. 2013), a system with similar out-of-plane spin-polarized valleys. The bottom of S_4 is detected as circular feature. The spin-asymmetry data provide proof of the out-of-plane but opposite spin polarization of S_3 and S_4. Evidently, around $\bar{K}$ and $\bar{K}'$, S_3 and S_4 form out-of-plane spin-polarized valleys with an almost parabolic shape.

At the $\bar{K}$ and $\bar{K}'$ points, S_3 and S_4 exhibit a giant spin-dependent splitting in energy of about 0.6 eV. This splitting is almost two thirds of the atomic spin-orbit splitting of a thallium p state ($\Delta E = (2l + 1)\lambda \approx 0.9$ for $l = 1$, $\lambda \approx 0.3$ eV), larger by more than a factor of two than observed in the occupied surface-state components S_1 and S_2 (Sakamoto et al. 2009). Charge distribution plots in Fig. 3.13a of the occupied and unoccupied surface-state components S_1 and S_3 provide an explanation. At $\bar{K}$ and $\bar{K}'$, the unoccupied state is mainly localized at the Tl atoms, whereas the occupied surface state is localized at the Si atoms. Charge distribution plots without considering spin-orbit interaction indicate similar results (Lee et al. 2002). The nature of the giant splitting is, therefore, assigned to the proximity of the unoccupied surface state to the heavy nuclei (Nagano et al. 2009). Enhanced spin-orbit-induced splittings may also be a result of an asymmetric charge distribution (Premper et al. 2007). In the case of Tl/Si(111)-(1×1), the k-integrated valence-band charge distribution, which enters the Poisson equation, is almost spherically symmetric close to the Tl atoms (see Fig. 3.13b) (Krüger 2013b). Therefore, this argument does not apply.

3.4.2 Spin Texture

Notably, at the $\bar{K}$ and $\bar{K}'$ points, the same features appear with opposite spin polarization (compare spectra in Fig. 3.14b, e). The spin-asymmetry data in Fig. 3.14c and f around $\bar{K}$ and $\bar{K}'$, respectively, underline this reversal. Spin-asymmetry values of more than 80 % around $\bar{K}$ and 60 % around $\bar{K}'$ are found (without background subtraction).

Figure 3.16b shows a plot of the out-of-plane spin-resolved spectral intensity at an energy of 0.2 eV as a function of the azimuth φ of the sample. If k_x and k_y are the components of $\mathbf{k}_\parallel$ in the x and y direction, respectively (see illustration in Fig. 3.16a), the measurement represents a momentum distribution curve with constant $k_\parallel = \sqrt{k_x^2 + k_y^2}$ but varying k_x and k_y components (see blue dashed line in Fig. 3.16a). At $\bar{K}$ and $\bar{K}'$, the measurement crosses the lower lying unoccupied surface-state component S_3. Clearly, S_3 is out-of-plane spin polarized antiparallel to the surface normal at $\bar{K}$ and out-of-plane spin polarized parallel to the surface normal at $\bar{K}'$. As discussed before, the different spectral intensities at $\bar{K}$ and $\bar{K}'$ are related to the threefold symmetry of the surface. Furthermore, the unpolarized background intensity can be easily identified. The presented results strongly indicate almost 100 % out-of-plane but oppositely spin-polarized valleys in the vicinity of E_F at $\bar{K}$ and $\bar{K}'$.

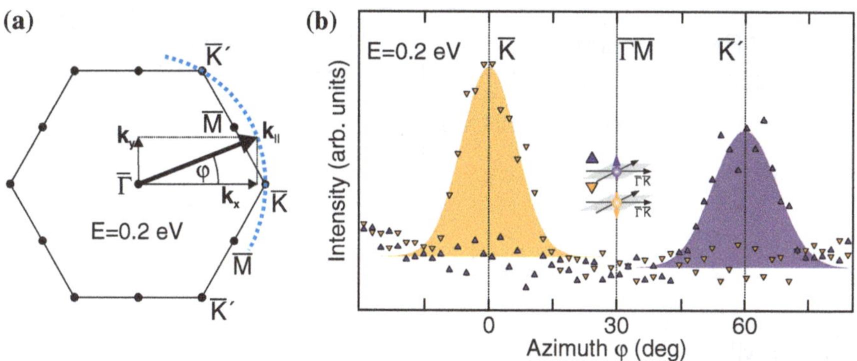

Fig. 3.16 **a** Illustration and **b** results of a constant-$k_{\|}$ ($k_{\|} \approx 1.1$ Å) measurement of Tl/Si(111)-(1×1) (preparation B) at a constant energy of $E = 0.2$ eV for $\theta \approx 60°$. $k_{\|} = \sqrt{k_x^2 + k_y^2}$ is fixed while k_x and k_y are varied by changing the sample azimuth $\varphi = \arctan \frac{k_y}{k_x}$. The measurement crosses the *lower* lying surface-state component S_3 at K̄ and K̄′. The experiment is sensitive to the out-of-plane spin-polarization direction (*purple* up- and *orange* down-pointing *triangles*). Note that the results were also used to determine the high-symmetry directions of the sample

3.4.3 Occupied Valleys

The unoccupied surface state found on the pristine Tl/Si(111)-(1×1) surface gives rise to almost completely out-of-plane spin-polarized valleys close to the Fermi level at the K̄ and K̄′ points. However, spintronic applications rely on metallic spin-polarized states. In the following, it will be demonstrated that the valleys become metallic by adding more Tl to the Tl/Si(111)-(1×1) surface. In this light, the unique properties of the Tl/Si(111)-(1×1) surface may open up an avenue to improve the efficiency of spin currents for silicon spintronics applications (Sakamoto et al. 2013).

Previous SARPES measurements detected occupied out-of-plane spin-polarized valleys for Tl/Si(111)-(1×1) surfaces with Tl coverages of more than 1 ML (Sakamoto et al. 2013). In Sakamoto et al. (2013), it is argued that the metallic valleys result from a downshift of the unoccupied surface-state component S_3 due to electron doping by the additional Tl atoms. Metallic valleys are also found for the isoelectronic Tl/Ge(111)-(1×1) surface (Ohtsubo et al. 2012). Here, it is assumed that the occupation of adsorption sites at surface defects by a small excess of Tl leads to electron doping mainly for the states localized in the topmost Tl layer.

ARPES and SR-IPE measurements on surfaces with additional Tl are presented in Fig. 3.17. Surfaces with Tl coverages exceeding 1 ML were prepared as described in Sect. 3.2.2. The Tl coverage has been determined by STM and by a comparison of the ARPES data obtained in this experiment with the SARPES data presented in Sakamoto et al. (2013). Tl coverages of (i) 1 ML, (ii) approximately 1.1 ML (see STM image in Sect. 3.2.2) and (iii) above 1.2 ML are analyzed. The SR-IPE and

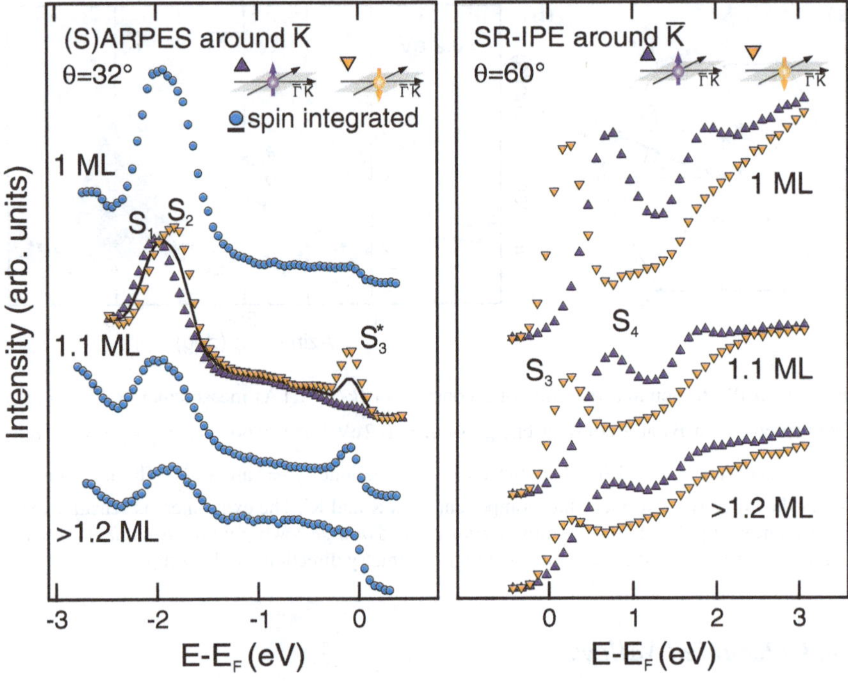

Fig. 3.17 (*left-hand side*) ARPES measurements ($\hbar\omega = 21.22\,\mathrm{eV}$) around $\bar{\mathrm{K}}$ of the 1 ML Tl/Si(111)-(1×1) surface (preparation B) and Tl/Si(111)-(1×1) surfaces with more than 1 ML Tl. The out-of-plane spin-resolved data are taken from Sakamoto et al. (2013) (with permission from K.S. Sakamoto). (*right-hand side*) Corresponding SR-IPE measurements with sensitivity to the out-of-plane spin-polarization direction

ARPES spectra were taken around the $\bar{\mathrm{K}}$ point. For each coverage, the experiments were conducted at identical samples.

(i) For 1 ML, the ARPES experiment reveals the occupied surface-state emissions S_1 and S_2. The SR-IPE spectra show the unoccupied surface-state components S_3 and S_4. Close to E_F, no occupied spectral feature is detected. A Fermi level cut-off is observed, which indicates a metallic-like character of the Tl/Si(111)-(1×1) surface. STM studies on Tl/Si(111)-(1 × 1) show that this may be a result of surface defects, which lead to an increased density of states near the Fermi level (Kocán et al. 2011).

(ii) At a coverage of about 1.1 ML, the spectral intensities of S_1 and S_2 decrease and an occupied state S_3^* appears at about 70 meV below the Fermi level. Good agreement is found in comparison with the SARPES data from Sakamoto et al. (2013) (see spin-resolved photoemission data in Fig. 3.17). S_3^* may be identified as the former unoccupied surface state S_3. This would correspond to a shift of S_3 of about 250 meV. However, no energy shift is observed for S_3 and S_4 in the corresponding SR-IPE data. Here, only a decrease of the spectral intensities is observed.

(iii) For coverages above 1.2 ML, S_3^* disappears. This is accompanied by a further decrease of the spectral intensities of S_1, S_2, S_3 and S_4. At the same time, the ARPES measurements show an increased background intensity in the vicinity of E_F, which indicates an increased amount of defects. Again, no shift of S_3 and S_4 is detected.

The results are interpreted as follows. The decrease of all spectral intensities S_1 to S_4 for coverages above 1 ML indicates a reduction of the surface quality. The higher the excess amount of Tl the higher the surface roughness and the lower the spectral intensities of the surface-state emissions. The transient coexistence of S_3 and S_3^* and the fact that no shift in energy is found for S_3 and S_4 are in contradiction with the interpretation of a homogenous doping. Rather, the SR-IPE results suggest the formation of patches on the Tl/Si(111)-(1×1) surface, where electron doping is effective. The SR-IPE results are also compatible with the interpretation that S_3 becomes partly occupied, due to a lifetime broadening as a result of a reduced surface quality.

Independent of the interpretation of S_3^*, Tl/Si(111)-(1×1) surfaces with a Tl coverage of about 1.1 ML give rise to metallic almost completely out-of-plane spin-polarized valleys at the $\bar{\text{K}}$ ($\bar{\text{K}}'$) points. No other states cross the Fermi level and, thus, spin transport is solely determined by the valleys. Remarkably, as has been demonstrated, the spin polarization is opposite at $\bar{\text{K}}$ and $\bar{\text{K}}'$. This leads to a Fermi surface as illustrated in Fig. 3.18, where scattering between the $\bar{\text{K}}$ and $\bar{\text{K}}'$ points is prohibited, since scattering due to nonmagnetic defects such as adatoms and step edges must conserve the spin polarization (Roushan et al. 2009; Hirayama et al. 2011). In combination with the high rigidity of the surface state (see Sect. 3.3.3), this may open a new perspective towards the realization of high efficient silicon spintronic devices (Sakamoto et al. 2013). Moreover, a resemblance to MoS_2 is found,

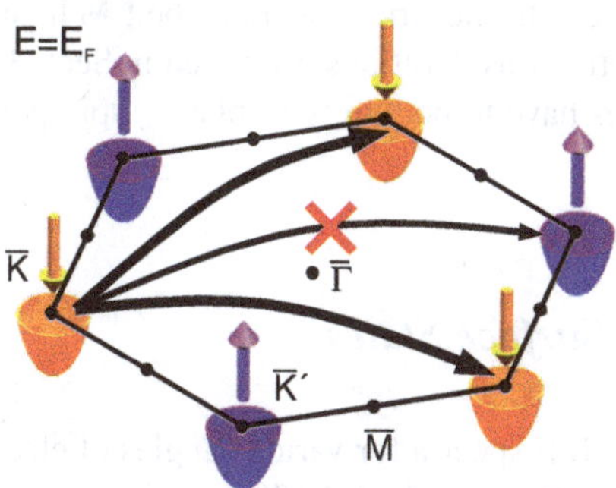

Fig. 3.18 Visualization of the Fermi surface of the doped Tl/Si(111)-(1×1) surface. Completely out-of-plane but oppositely spin-polarized valleys are formed at $\bar{\text{K}}$ and $\bar{\text{K}}'$, respectively. Scattering between valleys at $\bar{\text{K}}$ and $\bar{\text{K}}'$ is not allowed, as long as the spin is preserved in the scattering process (after Sakamoto et al. 2013)

a system which has attracted immense attention in the field of valleytronics (Xiao et al. 2012; Mak et al. 2012; Zeng et al. 2012; Feng et al. 2012). In both systems, the valleys at $\bar{\mathrm{K}}$ and $\bar{\mathrm{K}}'$ carry opposite Berry curvature. This allows for transverse currents where electrons with antipodal spin flow to opposite sides of the sample when an electrical field is applied in the surface plane.

3.4.4 Conclusion

In conclusion, at the $\bar{\mathrm{K}}$ ($\bar{\mathrm{K}}'$) points, the unoccupied surface state shows an extraordinary large splitting in energy of about 0.6 eV, which is traced back to the strong localization close to the Tl atoms. As a result, almost completely out-of-plane but oppositely spin-polarized valleys are formed in the vicinity of E_F at $\bar{\mathrm{K}}$ and $\bar{\mathrm{K}}'$. Furthermore, the completely out-of-plane spin-polarized valleys become metallic by adsorption of additional Tl. This gives rise to a peculiar Fermi surface, where backscattering is strongly suppressed.

3.5 Results for the $\bar{\Gamma}\bar{\mathrm{M}}$ Direction

So far, studies on the surface electronic structure of Tl/Si(111)-(1×1) concentrated on the $\bar{\Gamma}\bar{\mathrm{K}}$ ($\bar{\Gamma}\bar{\mathrm{K}}'$) directions, where occupied and unoccupied surface states with unique properties emerge: Rotating spin-polarization vectors and large energy splittings at the $\bar{\mathrm{K}}$ ($\bar{\mathrm{K}}'$) points. Investigations of the other high-symmetry directions have been limited to brief discussions in theoretical contributions (Ibañez-Azpiroz et al. 2011) and spin-integrated photoemission experiments (Lee et al. 2002; Sakamoto et al. 2009; Gruznev et al. 2014). The latter detect an occupied surface state along $\bar{\Gamma}\bar{\mathrm{M}}$, which becomes a surface-resonant state around $\bar{\mathrm{M}}$. In this section, the focus is put on the unoccupied surface electronic structure along the $\bar{\Gamma}\bar{\mathrm{M}}$ high-symmetry directions. Here, the basic symmetry considerations discussed in Sect. 3.2.2 demand that spin-orbit-split surface states have to be purely in-plane$_\perp$ spin polarized and degenerate at $\bar{\mathrm{M}}$.

3.5.1 Unoccupied Surface States

Figure 3.19a shows SR-IPE spectra for various angles of electron incidence θ along the $\bar{\Gamma}\bar{\mathrm{M}}$ high-symmetry direction. The SR-IPE experiment probes the in-plane$_\perp$ spin-polarization direction (red up- and blue down-pointing triangles) and a normalization to 100 % effective in-plane$_\perp$ beam polarization has been applied. The $E(\mathbf{k}_\parallel)$ plot in Fig. 3.20a is obtained by a peak analysis of the spectra (see Sect. 2.1), not all of which are shown in Fig. 3.19a (additional spectra are shown in Appendix A.1).

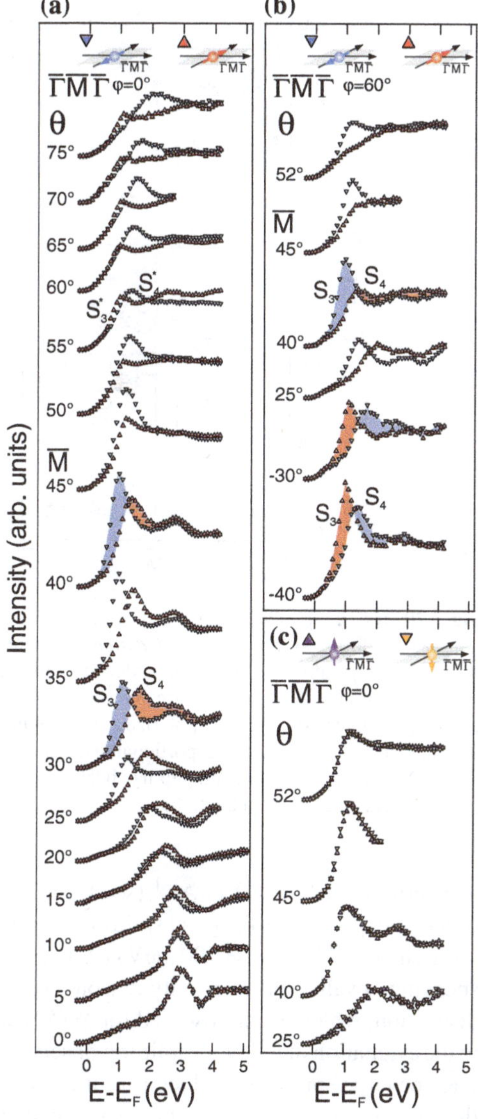

Fig. 3.19 **a** and **b** SR-IPE spectra of Tl/Si(111)-(1×1) (preparation A) along $\bar{\Gamma}\bar{M}$ with sensitivity to the in-plane$_\perp$ spin-polarization direction (*red up-* and *blue down-pointing triangles*). **c** SR-IPE spectra along $\bar{\Gamma}\bar{M}$ with sensitivity to the out-of-plane/in-plane$_\parallel$ spin-polarization direction (*purple up-pointing* and *orange down-pointing triangles*). θ and φ denote the angle of electron incidence and the sample azimuth, respectively

At $\bar{\Gamma}$, a spin-degenerate state is observed at about 2.8 eV (see also Stolwijk et al. 2013 and Sect. 3.3). For increasing momenta parallel to the surface, this state splits into two components, which are denoted as S_3 and S_4. For $\theta > 45°$, the experiment probes the second Brillouin zone. Here, the observed features are labeled S_3^* and S_4^*. $S_3(S_3^*)$ and $S_4(S_4^*)$ are in-plane$_\perp$ spin polarized. Notably, the Rashba splitting E_R, i.e., the difference between the crossing point and the band minimum, amounts to $E_R = 200$ meV. The momentum shift k_0 of the band extremum away from $\bar{M}$

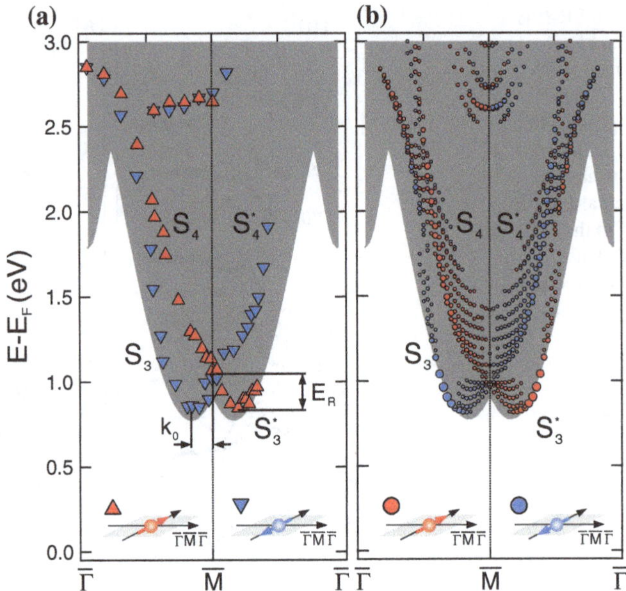

Fig. 3.20 **a** $E(\mathbf{k}_{\|})$ plot derived from spectra taken along $\bar{\Gamma}\bar{\mathrm{M}}$ with sensitivity to the in-plane$_{\perp}$ spin-polarization direction. The *gray-shaded area* represents the calculated projected bulk band structure from (**b**). To quantify the spin-orbit-induced splitting, the Rashba energy E_R and the shift k_0 of the band minimum away from $\bar{\mathrm{M}}$ are determined. **b** Calculated band structure including spin-orbit coupling (Krüger 2013c). Spin-polarized states within the surface region are marked by *circles*. The diameter of the *circles* is proportional to the spin polarization. Only in-plane$_{\perp}$ spin polarization exists. A correction of the band gap has been applied according to quasiparticle calculations and measurements presented in Sect. 3.3

is determined to $k_0 \approx 0.17\,\text{Å}^{-1}$ (see Fig. 3.20a and Sect. 1.2). This lies in the same order of magnitude as the giant splitting observed for the frequently discussed surface alloy Bi/Ag(111) ($E_R = 200\,\text{meV}$) (Ast et al. 2007). It should be noted that, strictly speaking, a value for α_R is only reasonable for a state with nearly free electron like dispersion, which is not observed for $S_3(S_3^*)$ and $S_4(S_4^*)$ (see also Sect. 3.7). Similar to the evaluation of the giant splitting observed for the unoccupied surface state along $\bar{\Gamma}\bar{\mathrm{K}}$, the large spin-orbit-induced splitting can be attributed to the close proximity of the surface-resonant state to the heavy nuclei of the Tl atoms.

A corresponding density functional theory calculation within the local-density approximation including spin-orbit interaction along $\bar{\Gamma}\bar{\mathrm{M}}$ is presented in Fig. 3.20b (Krüger 2013c).[6] According to the results of the quasiparticle calculation for Tl/Si(111) employing a small slab (Stolwijk et al. 2013) (see Fig. 3.4), the calculated eigenvalues are shifted by about 0.28 eV to compensate for the DFT band-gap

[6]Norm-conserving pseudopotentials that include scalar relativistic corrections and spin-orbit coupling have been employed together with a representation of the wave function by atom-centered Gaussian orbitals with s, p, d and s^* symmetry (Stärk et al. 2011). The Tl/Si(111) surface is treated

error. Spin-polarized states within the surface region are plotted as red or blues circles, where the diameter of the circles corresponds to the magnitude of the respective spin polarization. Only in-plane$_\perp$ spin polarization occurs. In comparison with the theoretical calculations (Krüger 2013c) in Fig. 3.20b, $S_3(S_3^*)$ and $S_4(S_4^*)$ are allocated to the unoccupied surface state of Tl/Si(111)-(1×1), which becomes a surface resonance along $\bar{\Gamma}\bar{M}$.

3.5.2 Spin Texture

At $\bar{M}$ ($\theta \approx 45°$, $k_\parallel \approx 0.95\,\text{Å}^{-1}$), S_3 and S_4 are energetically degenerate. However, the SR-IPE measurements detect a high intensity asymmetry of the two in-plane$_\perp$-polarized spin components (see Fig. 3.19a, b). In fact, this intensity asymmetry is also observed from $\theta = 30°$ up to $\theta = 75°$. As a rule, S_4 exhibits a lower spectral intensity than S_3. In the second Brillouin zone, S_3^* appears less intense than S_4^*. Figure 3.19c presents SR-IPE measurements with the spin sensitivity set to the out-of-plane/in-plane$_\parallel$ spin-polarization direction. Note that in these measurements, the intensity asymmetries at $\bar{M}$ and along $\bar{\Gamma}\bar{M}$ almost vanish. This indicates that the intensity asymmetry is related to the spin quantization axis and possible explanations such as the limited energy and momentum resolution of the experiment can be ruled out. Instead, the observations are explained in terms of spin-dependent transition probabilities. The off-normal direction of photon detection and the off-normal angle of electron incidence break the symmetry of the setup. This may, in analogy to photoemission for nonnormal photon incidence, lead to the observation of spin polarization, although the final state (initial state in photoemission) exhibits no spin polarization (Henk et al. 1996; Jozwiak et al. 2011; Heinzmann and Dil 2012; Park and Louie 2012; Jozwiak et al. 2013).

A closer investigation of the spin texture of the spin-orbit-split surface-resonant state is conducted with the help of further SR-IPE measurements. Figure 3.19b presents SR-IPE spectra taken with an azimuthal angle of the sample of $\varphi = 60°$, i.e., the neighboring $\bar{\Gamma}\bar{M}$ direction is measured. No change of the in-plane$_\perp$ spin-polarization direction of $S_3(S_3^*)$ and $S_4(S_4^*)$ is observed. The lower spectral intensity is attributed to the influence of the threefold symmetry. SR-IPE spectra with negative angles of electron incidence θ (see Fig. 3.19b) indicate a reversed spin polarization for S_3 and S_4. SR-IPE measurements with the spin sensitivity set to the out-of-plane/in-plane$_\parallel$ spin-polarization direction (see Fig. 3.19c) show no out-of-plane/in-plane$_\parallel$ spin polarization. In total, a clear indication for a Rashba-type spin splitting is at hand.

To further test the out-of-plane/in-plane$_\parallel$ spin degeneracy, Fig. 3.21b presents measurements for $\theta = 40°$ along $\mathbf{k}_\parallel$ directions at an angle of $\varphi = 8°$ ($\bar{\Gamma}\bar{\mu}^{8°}$) and

(Footnote 6 continued)
within a supercell approach. Around $\bar{M}$ point and along $\bar{\Gamma}\bar{M}$, the surface states are resonant with Si bulk states. For a proper representation of these surface resonances, a large slab with 70 Si substrate layers, a Tl adlayer and a hydrogen layer terminating the lower surface has been used.

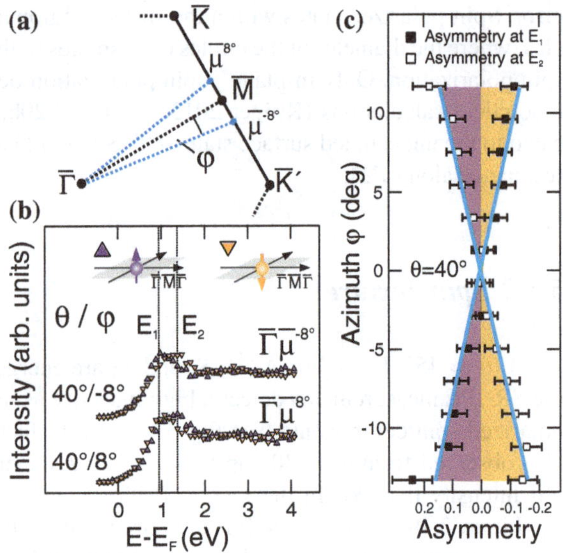

Fig. 3.21 **a** Illustration of the measured $\mathbf{k}_{\|}$ directions at an angle φ with respect to $\bar{\Gamma}\bar{\text{M}}$. **b** SR-IPE measurements of Tl/Si(111)-(1×1) (preparation A) with sensitivity to the out-of-plane/in-plane$_{\|}$ spin-polarization direction along directions at an angle $\varphi = 8°$ ($\bar{\Gamma}\bar{\mu}^{8°}$) and $\varphi = -8°$ ($\bar{\Gamma}\bar{\mu}^{-8°}$). **c** Plot of the observed spin asymmetry at E_1 and E_2 (see **b**) for $\theta = 40°$ as a function of the sample azimuth φ

$\varphi = -8°$ ($\bar{\Gamma}\bar{\mu}^{-8°}$) with respect to $\bar{\Gamma}\bar{\text{M}}$ (see blue lines in Fig. 3.21a). Along both directions, features with out-of-plane/in-plane$_{\|}$ spin polarization are found, whereas the spin polarization is reversed between $\bar{\Gamma}\bar{\mu}^{8°}$ and $\bar{\Gamma}\bar{\mu}^{-8°}$. To systematically evaluate the appearance of out-of-plane/in-plane$_{\|}$ spin polarization, the spin asymmetry is monitored at two discrete energies E_1 and E_2, while varying the azimuth φ of the sample (see Fig. 3.21c). E_1 and E_2 are chosen according to Fig. 3.21b, because at E_1 and E_2 large spin asymmetry is detected, which is oppositely oriented at E_1 and E_2. At E_1, the observed spin asymmetry decreases (increases) linearly up to $\varphi \approx 12°$ ($\varphi \approx -12°$). The opposite behavior is found at E_2. Remarkably, out-of-plane/in-plane$_{\|}$ spin asymmetry already emerges for $\varphi \approx 2°$ (see $\varphi \approx 2°$ in Fig. 3.21c), which corresponds to a shift of about 0.03 Å^{-1} away from $\bar{\Gamma}\bar{\text{M}}$. Ultimately, the results demonstrate the relevance of the symmetry of the system, i.e., the mirror plane along $\bar{\Gamma}\bar{\text{M}}$, for the spin textures of the surface states.

3.5.3 Conclusion

In conclusion, along $\bar{\Gamma}\bar{\text{M}}$, we detected a surface-resonant state with giant spin-orbit-induced spin splitting, which has not been identified in literature, so far, neither experimentally nor theoretically. Evidently, along $\bar{\Gamma}\bar{\text{M}}$, the spin-polarization direction is confined to the in-plane$_{\perp}$ spin-polarization direction. This is in agreement with the symmetry of the surface, as the $\bar{\Gamma}\bar{M}$ directions lie in the mirror planes of the system. Intriguingly, this restriction is limited to the high-symmetry line, as additional spin-polarization directions emerge already for small deviations from the mirror plane. A fact, which can be used as a sensor to calibrate the $\bar{\Gamma}\bar{M}$ direction, but also stresses the importance of a correct sample and electron beam alignment to avoid misinterpretations of experimental data, when investigating complex k-dependent spin textures.

3.6 Results for the $\bar{K}\bar{M}(\bar{K}'\bar{M})$ Direction

For Tl/Si(111)-(1×1), an unoccupied spin-orbit-split surface state is found along $\bar{\Gamma}\bar{K}$ and $\bar{\Gamma}\bar{M}$. This section presents the results for the $\bar{K}\bar{M}$ ($\bar{K}'\bar{M}$) direction. Here, band structure calculations already predict a state, which can be viewed as the continuation of the surface state observed along $\bar{\Gamma}\bar{K}$ (see Sect. 3.1).

3.6.1 Unoccupied Surface States

The spin-dependent surface electronic structure along the $\bar{K}\bar{M}$ ($\bar{K}'\bar{M}$) direction is accessed in two ways:

(i) SR-IPE measurements along $\mathbf{k}_{\parallel}$ directions in between $\bar{\Gamma}\bar{M}$ and $\bar{\Gamma}\bar{K}$ ($\bar{\Gamma}\bar{K}'$) with sensitivity to the out-of-plane/in-plane$_{\parallel}$ spin-polarization direction. Figure 3.22b shows a sketch of the measured directions denoted as $\bar{\Gamma}\bar{\mu}_{1...13}$. The measurement direction is set by varying the azimuth φ of the sample. Note that the SR-IPE experiment is sensitive to the out-of-plane and in-plane$_{\parallel}$ spin-polarization direction with respect to $\mathbf{k}_{\parallel}$. As the measurement directions intersect the $\bar{K}\bar{M}$ ($\bar{K}'\bar{M}$) line (see Fig. 3.22c), the experiment simultaneously probes the out-of-plane and in-plane$_{\perp}$ spin-polarization direction with respect to $\bar{K}\bar{M}$ ($\bar{K}'\bar{M}$). It will be demonstrated that the out-of-plane spin polarization is dominant. Therefore, spectra have been normalized to 100 % out-of-plane spin polarization (purple up- and orange down-pointing triangles). The obtained spectra are presented in Fig. 3.22a, d. The angle of electron incidence θ and the azimuth of the sample φ are set, so that the lower lying state S_3 is measured along $\bar{K}\bar{M}$ ($\bar{K}'\bar{M}$). As a result, S_4 is measured at a slightly shifted position $\Delta\mathbf{k}$ in k space ($\Delta k \leq 0.05\,\text{Å}^{-1}$).
(ii) SR-IPE measurements along $\bar{\Gamma}\bar{K}\bar{M}$ and $\bar{\Gamma}\bar{K}'\bar{M}$ with sensitivity to the out-of-plane/in-plane$_{\parallel}$ spin-polarization direction, as presented in Sect. 3.3. Along $\bar{\Gamma}\bar{K}\bar{M}$, no in-plane$_{\parallel}$ spin polarization occurs (Stolwijk et al. 2013). The experiment detects, therefore, only the out-of-plane spin-polarization component.

Along $\bar{K}\bar{M}$ ($\bar{K}'\bar{M}$), two dispersive features are identified: S_3 and S_4 (see Fig. 3.22). The corresponding $E(\mathbf{k}_{\parallel})$ plot in Fig. 3.23a is obtained by a peak analysis (see Sect. 2.1) of the spectra presented in Figs. 3.22 and 3.7 and gives an overview of the dispersion of S_3 and S_4. The dispersion agrees well with the theoretical

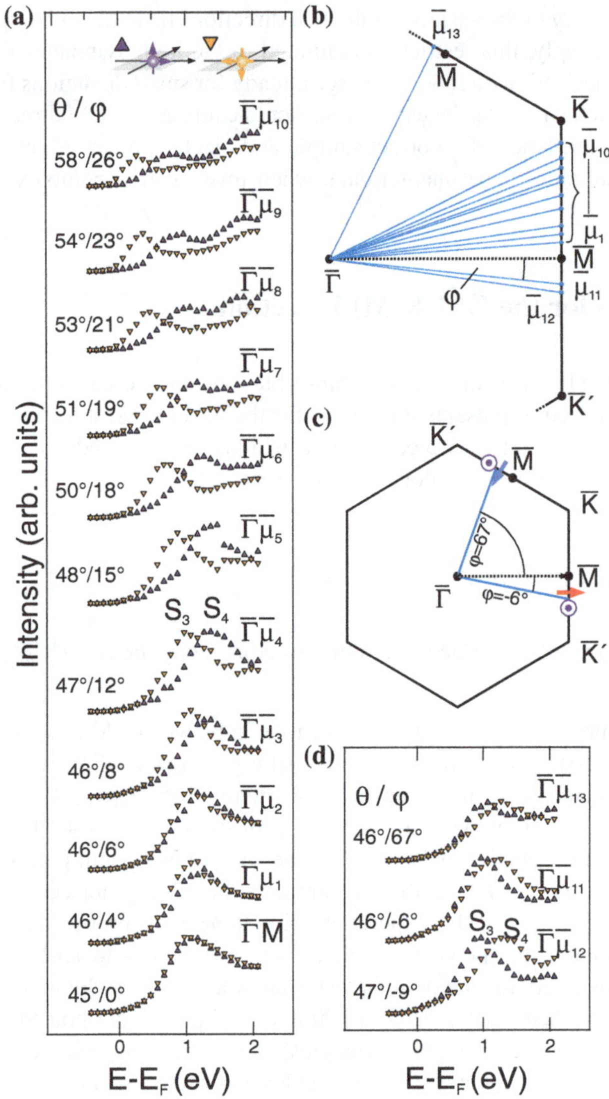

Fig. 3.22 **a, d** SR-IPE spectra of Tl/Si(111)-(1×1) (preparation A) along $\mathbf{k}_{\|}$ directions in between $\bar{\Gamma}\bar{\text{M}}$ and $\bar{\Gamma}\bar{\text{K}}(\bar{\Gamma}\bar{\text{K}}')$ with sensitivity to the out-of-plane/in-plane$_{\|}$ spin-polarization direction. **b** Illustration of the conducted SR-IPE measurements along $\mathbf{k}_{\|}$ directions in between $\bar{\Gamma}\bar{\text{M}}$ and $\bar{\Gamma}\bar{\text{K}}$ ($\bar{\Gamma}\bar{\text{K}}'$), denoted as $\bar{\Gamma}\bar{\mu}_{1...13}$. **c** Illustration of the spin-polarization components of S_3 according to the calculations in Fig. 3.23

findings (see Fig. 3.23b, c) (Krüger 2013c). Around $\bar{\text{M}}$, S_3 and S_4 are degenerate (see $\theta = 45°$ in Fig. 3.22a). Approaching $\bar{\text{K}}$, S_3 and S_4 become spin split and exhibit out-of-plane/in-plane$_{\|}$ spin polarization. Around $\bar{\text{K}}(\bar{\text{K}}')$, S_3 and S_4 form valleys,

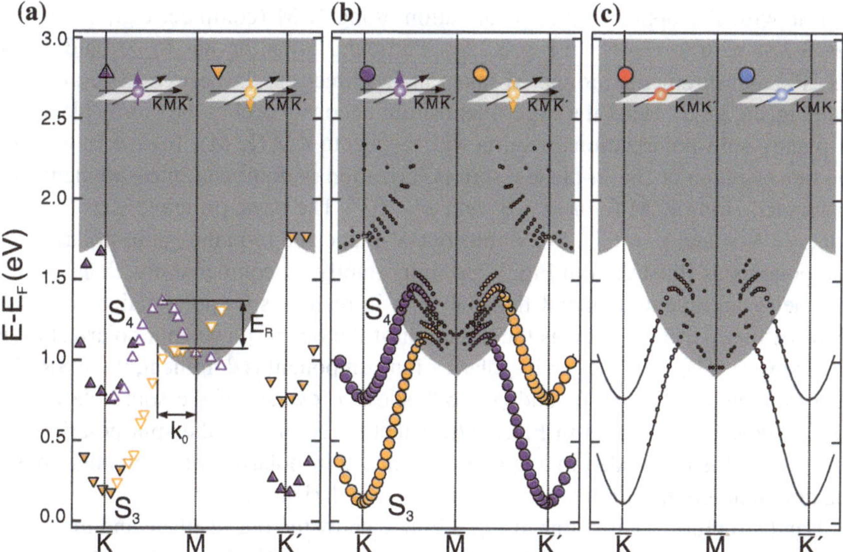

Fig. 3.23 **a** $E(\mathbf{k}_{\|})$ plot derived from spectra shown in Fig. 3.22 (*open triangles*) and Figs. 3.7 and 3.10 (*filled triangles*). The *gray-shaded area* represents the calculated projected bulk band structure. The Rashba energy E_R and the momentum shift k_0 are determined to $E_R = 320$ meV and $k_0 = 0.20\text{Å}^{-1}$. **b** and **c** present band structure calculations within the local-density approximation including spin-orbit interaction along $\bar{K}\bar{M}\bar{K}'$ illustrating **b** the out-of-plane and **c** in-plane$_{\perp}$ spin polarization as *circles* (Krüger 2013c). The diameter of the *circles* is proportional to the spin polarization, with a maximum degree of 100 %, e.g., S_4 at $\bar{K}$. A correction of the band gap has been applied according to quasiparticle calculations and measurements presented in Sect. 3.3 (see also discussion in Sect. 3.5.1)

which are split in energy by about 0.6 eV. In comparison with the results for the $\bar{\Gamma}\bar{K}$ ($\bar{\Gamma}\bar{K}'$) direction (see Sect. 3.3), S_3 and S_4 are easily identified as the continuation of the surface-state components observed along $\bar{\Gamma}\bar{K}$ ($\bar{\Gamma}\bar{K}'$). These become surface resonances around $\bar{M}$. A giant spin splitting with $E_R = 320$ meV and $k_0 = 0.20\,\text{Å}^{-1}$ is detected (see Fig. 3.23a). This even exceeds the giant splitting observed for the Bi/Ag(111) alloy. It is reasonable to attribute the large spin splitting to the localization of the state close to the Tl atoms.

3.6.2 Spin Texture

In the following, the spin texture of S_3 and S_4 is addressed. At $\bar{M}$, the experimental results demonstrate that S_3 and S_4 are energetically degenerate and exhibit no out-of-plane/in-plane$_{\|}$ spin polarization (see $\theta = 45°$ in Fig. 3.22a and Sect. 3.5.2). Furthermore, around $\bar{K}$ ($\bar{K}'$), S_3 and S_4 are almost completely out-of-plane spin polarized (see $\theta = 55°$–$70°$ in Fig. 3.7). Compared to the $\bar{K}\bar{M}$ direction, S_3 and S_4

appear with the opposite spin polarization along $\bar{\text{K}}'\bar{\text{M}}$ (compare, e.g., $\varphi = 6°$ in Fig. 3.22a with $\varphi = -6°$ in Fig. 3.22d). Particular attention has to be paid to the SR-IPE measurements around $\bar{\text{M}}$ taken at $\mathbf{k}_{\parallel}$ directions in between $\bar{\Gamma}\bar{\text{K}}$ and $\bar{\Gamma}\bar{\text{M}}$ (see Fig. 3.22a, d). Here, the experiment simultaneously probes the out-of-plane and in-plane$_{\perp}$ spin-polarization direction with respect to $\bar{\text{K}}\bar{\text{M}}$ ($\bar{\text{K}}'\bar{\text{M}}$). To discern whether the out-of-plane or the in-plane$_{\perp}$ spin polarization is dominant, measurements are conducted along $\bar{\text{K}}'\bar{\text{M}}$ for $\varphi = -6°$ and $\varphi = 67°$. The basic principle is as follows. For $\varphi = -6°$ and $\varphi = 67°$, S_3 exhibits out-of-plane and in-plane$_{\perp}$ spin-polarization components as illustrated in Fig. 3.23c. Importantly, in comparison with $\varphi = -6°$, symmetry arguments demand that only the in-plane$_{\perp}$ spin-polarization direction reverses for $\varphi = 67°$, whereas the direction of the out-of-plane spin polarization is not changed. The in-plane$_{\perp}$ spin-polarization component is dominant, if the SR-IPE measurements for $\varphi = -6°$ and $\varphi = 67°$ show a reversal of the spin polarization of S_3. Clearly, the results in Fig. 3.22d detect no change of the spin polarization. Therefore, it is concluded that the out-of-plane spin-polarization component poses the dominant contribution to the observed spin polarization.

Band structure calculations in Fig. 3.23b, c show the out-of-plane and in-plane$_{\perp}$ spin-polarization components of S_3 and S_4, respectively (Krüger 2013c). Along $\bar{\text{K}}\bar{\text{M}}$ ($\bar{\text{K}}'\bar{\text{M}}$), the spin texture contains out-of-plane and in-plane$_{\perp}$ spin-polarization components. In-plane$_{\parallel}$ spin polarization is not allowed as a consequence of symmetry arguments. At $\bar{\text{K}}$ ($\bar{\text{K}}'$), S_3 and S_4 are almost completely out-of-plane spin polarized. Theory shows that approaching $\bar{\text{M}}$, the in-plane$_{\perp}$ spin-polarization component increases but is always at least a factor of about two smaller than the out-of-plane spin-polarization component. At $\bar{\text{M}}$, the surface state is degenerate. The SR-IPE experiments verify these findings.

3.6.3 Conclusion

In conclusion, along the $\bar{\text{K}}\bar{\text{M}}$ ($\bar{\text{K}}'\bar{\text{M}}$) high-symmetry direction, a spin-orbit-split surface state is detected, which can be viewed as the continuation of the unoccupied surface state found along $\bar{\Gamma}\bar{\text{K}}$ (see, e.g., band structure calculations in Sect. 3.1). In the vicinity of $\bar{\text{M}}$, the surface state possesses a giant spin splitting and is predominately out-of-plane spin polarized. Interestingly, this is in contrast to the $\bar{\Gamma}\bar{\text{M}}$ direction, where the surface state is purely in-plane spin polarized. This gives rise to a complex spin texture around $\bar{\text{M}}$, which is evaluated in the next section.

3.7 Results Around $\bar{\text{M}}$

The investigations of Tl/Si(111)-(1×1) unveil spin-orbit-split surface states along $\bar{\Gamma}\bar{\text{M}}$ and $\bar{\text{K}}\bar{\text{M}}$ ($\bar{\text{K}}'\bar{\text{M}}$), which exhibit giant spin splittings around $\bar{\text{M}}$. Along $\bar{\Gamma}\bar{\text{M}}$, only in-plane$_{\perp}$ spin polarization occurs, whereas along $\bar{\text{K}}\bar{\text{M}}$ ($\bar{\text{K}}\bar{\text{M}}'$) out-of-plane spin polarization is dominant. The findings give rise to the question that is at the center of this section. How is the spin texture around $\bar{\text{M}}$?

3.7.1 Spin Texture

The first step is to understand the dispersion behavior of the surface state around $\bar{\text{M}}$. For this purpose, the band structure is modeled with the help of a simple approach using an effective Hamiltonian H_{eff}, which takes into account the C_{1h} symmetry of the $\bar{\text{M}}$ point (mirror plane along $\mathbf{k_x}$) and first order Rashba parameters α (Vajna et al. 2012):

$$H_{\text{eff}} = H(\mathbf{k_x}) + H(\mathbf{k_y}) + H_R, \tag{3.1}$$

$$H(\mathbf{k_x}) = \frac{\hbar^2}{2m^*_{x4}} k_x^4, \tag{3.2}$$

$$H(\mathbf{k_y}) = \frac{\hbar^2}{2m^*_{y2}} k_y^2 + \frac{\hbar^2}{2m^*_{y4}} k_y^4 + \frac{\hbar^2}{2m^*_{y6}} k_y^6, \tag{3.3}$$

$$H_R = \alpha_1 k_x \sigma_y + \alpha_2 k_y \sigma_x + \alpha_3 k_y \sigma_z. \tag{3.4}$$

The C_{1h} symmetry implies different dispersion behavior along $\bar{\text{M}}\bar{\Gamma}$ ($\mathbf{k_x}$) and $\bar{\text{M}}\bar{\text{K}}$ ($\bar{\text{M}}\bar{\text{K}}'$) ($\mathbf{k_y}$). The essential parameters for $H(\mathbf{k_x})$ and $H(\mathbf{k_y})$ are deduced from fits (red dotted lines in Fig. 3.24) to the experimental spin-integrated $E(\mathbf{k}_\parallel)$ dispersion data (black markers in Fig. 3.24). The dispersion along the $\bar{\text{M}}\bar{\Gamma}$ direction is described by a polynomial of the order k_x^4 with an effective mass of $m^*_{x4} \approx 0.2\, m_e$. Along $\bar{\text{M}}\bar{\text{K}}$, a good approximation is found by taking into account contributions up to the order

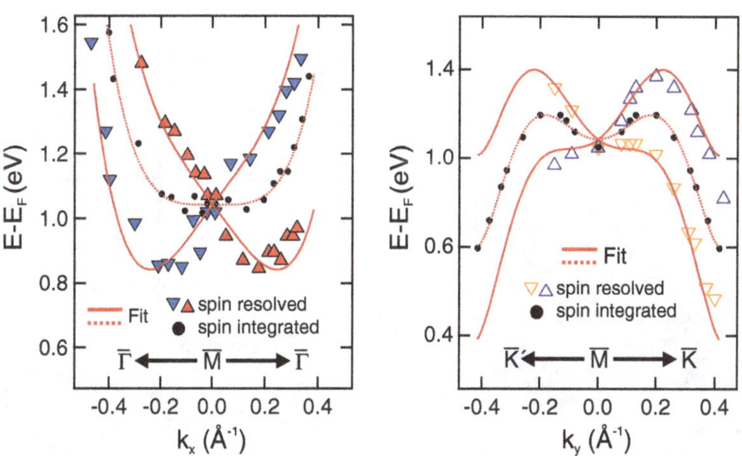

Fig. 3.24 $E(\mathbf{k}_\parallel)$ results obtained by SR-IPE around $\bar{\text{M}}$ along (*left-hand side*) $\bar{\text{M}}\bar{\Gamma}$ and (*right-hand side*) $\bar{\text{M}}\bar{\text{K}}$ ($\bar{\text{M}}\bar{\text{K}}'$) from Sects. 3.5 and 3.6, respectively. The *red dotted lines* represent fits to the spin-integrated data (*black markers*). *Red solid lines* denote fits to the spin-resolved data

k_y^6 with $m_{y2}^* \approx 0.55\,m_e$, $m_{y4}^* \approx -0.032\,m_e$ and $m_{y6}^* \approx 0.010\,m_e$. According to the model, the spin-resolved dispersion is given by

$$E_\pm = E_0 + \frac{\hbar^2 k_x^4}{2m_{x4}^*} + \frac{\hbar^2 k_y^2}{2m_{y2}^*} + \frac{\hbar^2 k_y^4}{2m_{y4}^*} + \frac{\hbar^2 k_y^6}{2m_{y6}^*} \pm \sqrt{\alpha_1^2 k_x^2 + \alpha_2^2 k_y^2 + \alpha_3^2 k_y^2}, \quad (3.5)$$

where E_0 is the crossing point of the two spin-split bands and the spin polarization is

$$\mathbf{P} = (p_x, p_y, p_z) = \frac{(\alpha_2 k_y, \alpha_1 k_x, \alpha_3 k_y)}{\left|(\alpha_2 k_y, \alpha_1 k_x, \alpha_3 k_y)\right|}. \quad (3.6)$$

The Rashba parameter $\alpha_1 \approx -1\,\text{eVÅ}$ is determined by fitting the spin-resolved $E(\mathbf{k}_\parallel)$ data along $\bar{\text{M}}\bar{\Gamma}$ ($k_y = 0$) (red lines in Fig. 3.24). Along $\bar{\text{K}}\bar{\text{M}}$ ($k_x = 0$), the SR-IPE data contain information on the out-of-plane and in-plane$_\perp$ spin-polarization components (see Sect. 3.6). Therefore, separate values for α_2 and α_3, which relate to the in-plane$_\perp$ (p_x) and out-of-plane (p_z) spin-polarization components, respectively, can not be experimentally obtained. Instead, the fit yields $\overline{\alpha} = \sqrt{\alpha_2^2 + \alpha_3^2} \approx 0.95\,\text{eVÅ}$. Values for α_2 and α_3 are estimated by a comparison with theory, which predict a ratio of $\frac{p_x}{p_z} \approx 0.5$ in the vicinity of $\bar{\text{M}}$. Accordingly, one obtains $\alpha_2 \approx 0.43\,\text{eVÅ}$ and $\alpha_3 \approx 0.85\,\text{eVÅ}$.

Figure 3.25a shows the modeled band dispersion. Around $\bar{\text{M}}$, the surface-state band E_+ forms a cone-like feature. Figure 3.25b shows the contour of this state (blue

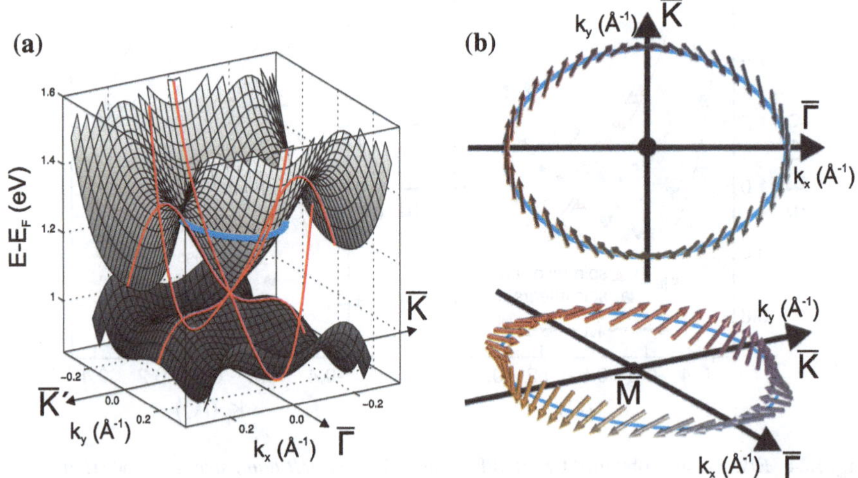

Fig. 3.25 **a** Illustration of the modeled $E(\mathbf{k}_\parallel)$ dispersion around $\bar{\text{M}}$. The *red lines* represent the fit results along $\bar{\text{M}}\bar{\Gamma}$ ($\mathbf{k_x}$) and $\bar{\text{M}}\bar{\text{K}}$ ($\bar{\text{M}}\bar{\text{K}}'$) ($\mathbf{k_y}$). **b** The *blue line* shows the (*top*) *top view* and (*bottom*) *tilted view* of the constant-energy contour (k_x-k_y plane) of the surface-state component E_+ at $E = 1.3\,\text{eV}$ (see also *blue line* in (**a**)). The spin texture is illustrated by *arrows*

line) in the k_x-k_y plane (top view and tilted view) at an energy of 1.3 eV. The surface-state component exhibits an elliptical shape in accordance with the C_{1h} symmetry of the $\bar{\text{M}}$ point. Note that systems with C_{2v} symmetry, such as W(110) and Au(110), possess similar anisotropic dispersions (Simon et al. 2010; Miyamoto et al. 2012). However, in contrast to systems with C_{2v} symmetry, the spin texture may contain out-of-plane spin-polarization components. The spin texture deduced from the model is illustrated for the constant-energy contour at 1.3 eV (see arrows in Fig. 3.25b). At the $\bar{\text{M}}\bar{\Gamma}$ intersections, the state is completely in-plane$_\perp$ but oppositely spin polarized. On the other hand, at the $\bar{\text{M}}\bar{\text{K}}$ and $\bar{\text{M}}\bar{\text{K}}'$ intersections, the spin-polarization direction is predominately antiparallel and parallel to the surface normal, respectively. The spin texture can be seen as a spin chirality in reciprocal space around $\bar{\text{M}}$.

3.7.2 Conclusion

Intriguingly, the results obtained for $\bar{\text{M}}\bar{\text{K}}$ and $\bar{\Gamma}\bar{\text{M}}$ indicate a peculiar spin texture around $\bar{\text{M}}$. Along $\bar{\text{M}}\bar{\text{K}}$, the surface state is predominately out-of-plane spin polarized, whereas it is purely in-plane spin polarized along $\bar{\Gamma}\bar{\text{M}}$. Light is shed onto the spin texture by fitting an effective Hamiltonian to the experimental data. This yields a model of the $E(\mathbf{k}_{\parallel})$ dispersion around $\bar{\text{M}}$ and a direct way to model the spin polarization. At a given energy, the $\bar{\text{M}}$ point is encircled by a surface-state component with a spin texture, which can be viewed as spin chirality in momentum space.

References

Altmann, W., Donath, M., Dose, V., Goldmann, A.: Dispersion of empty surface states on Ni(110). Solid State Commun. **53**, 209 (1985)

Ast, C.R., Henk, J., Ernst, A., Moreschini, L., Falub, M.C., Pacilé, D., Bruno, P., Kern, K., Grioni, M.: Giant spin splitting through surface alloying. Phys. Rev. Lett. **98**, 186807 (2007)

Awschalom, D.D., Flatte, M.E.: Challenges for semiconductor spintronics. Nat. Phys. **3**, 153 (2007)

Barke, I., Zheng, F., Rügheimer, T.K., Himpsel, F.J.: Experimental evidence for spin-split bands in a one-dimensional chain structure. Phys. Rev. Lett. **97**, 226405 (2006)

Basak, S., Lin, H., Wray, L.A., Xu, S.Y., Fu, L., Hasan, M.Z., Bansil, A.: Spin texture on the warped Dirac-cone surface states in topological insulators. Phys. Rev. B **84**, 121401 (2011)

Bondarenko, L.V., Gruznev, D.V., Yakovlev, A.A., Tupchaya, A.Y., Usachov, D., Vilkov, O., Fedorov, A., Vyalikh, D.V., Eremeev, S.V., Chulkov, E.V., Zotov, A.V., Saranin, A.A.: Large spin splitting of metallic surface-state bands at adsorbate-modified gold/silicon surfaces. Sci. Rep. **3**, 1826 (2013)

Bychkov, Y.A., Rashba, E.I.: Properties of a 2D electron gas with lifted spectral degeneracy. JETP Lett. **39**, 78 (1984)

Castellarin-Cudia, C., Surnev, S., Ramsey, M.G., Netzer, F.P.: Surface structures of thallium on Ge(111). Surf. Sci. **491**, 29 (2001)

Cheng, C., Kunc, K.: Structure and stability of Bi layers on Si(111) and Ge(111) surfaces. Phys. Rev. B **56**, 10283 (1997)

Datta, S., Das, B.: Electronic analog of the electro-optic modulator. Appl. Phys. Lett. **56**, 665 (1990)

Dil, J.H., Meier, F., Lobo-Checa, J., Patthey, L., Bihlmayer, G., Osterwalder, J.: Rashba-type spin-orbit splitting of quantum well states in ultrathin Pb films. Phys. Rev. Lett. **101**, 266802 (2008)

Dose, V., Fauster, T., Schneider, R.: Improved resolution in VUV isochromat spectroscopy. Appl. Phys. A Mater. Sci. Process. **40**, 203 (1986)

Dyakonov, M.I. (ed.): Spin Physics in Semiconductors. Springer, Berlin (2008)

Fabian, J., Matos-Abiague, A., Ertler, C., Stano, P., Žutić, I.: Semiconductor spintronics. Acta Phys. Slovaca **57**, 565 (2007)

Feng, W., Yao, Y., Zhu, W., Zhou, J., Yao, W., Xiao, D.: Intrinsic spin Hall effect in monolayers of group-VI dichalcogenides: a first-principles study. Phys. Rev. B **86**, 165108 (2012)

Gierz, I., Suzuki, T., Frantzeskakis, E., Pons, S., Ostanin, S., Ernst, A., Henk, J., Grioni, M., Kern, K., Ast, C.R.: Silicon surface with giant spin splitting. Phys. Rev. Lett. **103**, 046803 (2009)

Greenwood, N.N., Earnshaw, A. (eds.): Chemistry of the Elements. Pergamon Press, Oxford (1984)

Gruznev, D.V., Bondarenko, L.V., Matetskiy, A.V., Yakovlev, A.A., Tupchaya, A.Y., Eremeev, S.V., Chulkov, E.V., Chou, J.P., Wei, C.M., Lai, M.Y., Wang, Y.L., Zotov, A.V., Saranin, A.A.: A strategy to create spin-split metallic bands on silicon using a dense alloy layer. Sci. Rep. **4**, 4742 (2014)

Hamers, R.J., Tromp, R.M., Demuth, J.E.: Surface electronic structure of Si(111)-(7×7) resolved in real space. Phys. Rev. Lett. **56**, 1972 (1986)

Hatta, S., Aruga, T., Ohtsubo, Y., Okuyama, H.: Large Rashba spin splitting of surface resonance bands on semiconductor surface. Phys. Rev. B **80**, 113309 (2009)

Hatta, S., Kato, C., Tsuboi, N., Takahashi, S., Okuyama, H., Aruga, T., Harasawa, A., Okuda, T., Kinoshita, T.: Atomic and electronic structure of Tl/Ge(111)-(1×1) : LEED and ARPES measurements and first-principles calculations. Phys. Rev. B **76**, 075427 (2007)

Heinzmann, U., Dil, J.H.: Spin-orbit-induced photoelectron spin polarization in angle-resolved photoemission from both atomic and condensed matter targets. J. Phys. Condens. Matter **24**, 173001 (2012)

Henk, J., Scheunemann, T., Halilov, S.V., Feder, R.: Magnetic dichroism and electron spin polarization in photoemission: analytical results. J. Phys. Condens. Matter **8**, 47 (1996)

Herdt, A., Plucinski, L., Bihlmayer, G., Mussler, G., Döring, S., Krumrain, J., Grützmacher, D., Blügel, S., Schneider, C.M.: Spin-polarization limit in Bi_2Te_3 Dirac cone studied by angle- and spin-resolved photoemission experiments and ab initio calculations. Phys. Rev. B **87**, 035127 (2013)

Himpsel, F.J.: Inverse photoemission from semiconductors. Surf. Sci. Rep. **12**, 3 (1990)

Hirayama, H., Aoki, Y., Kato, C.: Quantum interference of Rashba-type spin-split surface state electrons. Phys. Rev. Lett. **107**, 027204 (2011)

Hochstrasser, M., Tobin, J.G., Rotenberg, E., Kevan, S.D.: Spin-resolved photoemission of surface states of W(110)-(1×1)H. Phys. Rev. Lett. **89**, 216802 (2002)

Hoesch, M., Muntwiler, M., Petrov, V.N., Hengsberger, M., Patthey, L., Shi, M., Falub, M., Greber, T., Osterwalder, J.: Spin structure of the Shockley surface state on Au(111). Phys. Rev. B **69**, 241401 (2004)

Höfer, U., Li, L., Ratzlaff, G.A., Heinz, T.F.: Nonlinear optical study of the Si(111)7×7 to 1×1 phase transition: superheating and the nature of the 1×1 phase. Phys. Rev. B **52**, 5264 (1995)

Höpfner, P., Schäfer, J., Fleszar, A., Dil, J.H., Slomski, B., Meier, F., Loho, C., Blumenstein, C., Patthey, L., Hanke, W., Claessen, R.: Three-dimensional spin rotations at the Fermi surface of a strongly spin-orbit coupled surface system. Phys. Rev. Lett. **108**, 186801 (2012)

Hwang, C., Kim, N., Lee, G., Shin, S., Uhm, S., Kim, H., Kim, J., Chung, J.: Origin of unusual work function change upon forming Tl nanoclusters on Si(111)-7×7 surface. Appl. Phys. A **89**, 431 (2007)

Hybertsen, M.S., Louie, S.G.: Model dielectric matrices for quasiparticle self-energy calculations. Phys. Rev. B **37**, 2733 (1988)

Ibañez-Azpiroz, J., Eiguren, A., Bergara, A.: Relativistic effects and fully spin-polarized Fermi surface at the Tl/Si(111) surface. Phys. Rev. B **84**, 125435 (2011)

Jia, J., Wang, J.Z., Liu, X., Xue, Q.K., Li, Z.Q., Kawazoe, Y., Zhang, S.B.: Artificial nanocluster crystal: lattice of identical Al clusters. Appl. Phys. Lett. **80**, 3186 (2002)

Jozwiak, C., Chen, Y.L., Fedorov, A.V., Analytis, J.G., Rotundu, C.R., Schmid, A.K., Denlinger, J.D., Chuang, Y.D., Lee, D.H., Fisher, I.R., Birgeneau, R.J., Shen, Z.X., Hussain, Z., Lanzara, A.: Widespread spin polarization effects in photoemission from topological insulators. Phys. Rev. B **84**, 165113 (2011)

Jozwiak, C., Park, C.H., Gotlieb, K., Hwang, C., Lee, D.H., Louie, S.G., Denlinger, J.D., Rotundu, C.R., Birgeneau, R.J., Hussain, Z., Lanzara, A.: Photoelectron spin-flipping and texture manipulation in a topological insulator. Nat. Phys. **9**, 293 (2013)

Kim, N.D., Hwang, C.G., Chung, J.W., Kim, T.C., Kim, H.J., Noh, D.Y.: Structural properties of a thallium-induced Si(111)-1$\times$1 surface. Phys. Rev. B **69**, 195311 (2004)

Kirchmann, P.S., Wolf, M., Dil, J.H., Horn, K., Bovensiepen, U.: Quantum size effects in Pb/Si(111) investigated by laser-induced photoemission. Phys. Rev. B **76**, 075406 (2007)

Kocán, P., Sobotík, P., Matvija, P., Setvín, M., Ošt'ádal, I.: An STM study of desorption-induced thallium structures on the Si(111) surface. Surf. Sci. **606**, 991 (2012)

Kocán, P., Sobotík, P., Ošt'ádal, I.: Metallic-like thallium overlayer on a Si(111) surface. Phys. Rev. B **84**, 233304 (2011)

Kormányos, A., Zólyomi, V., Drummond, N.D., Rakyta, P., Burkard, G., Fal'ko, V.I.: Monolayer MoS_2: trigonal warping, the γ valley, and spin-orbit coupling effects. Phys. Rev. B **88**, 045416 (2013)

Koroteev, Y.M., Bihlmayer, G., Gayone, J.E., Chulkov, E.V., Blügel, S., Echenique, P.M., Hofmann, P.: Strong spin-orbit splitting on Bi surfaces. Phys. Rev. Lett. **93**, 046403 (2004)

Kotlyar, V.G., Saranin, A.A., Zotov, A.V., Kasyanova, T.V.: Thallium overlayers on Si(111) studied by scanning tunneling microscopy. Surf. Sci. **543**, L663 (2003)

Kotlyar, V.G., Zotov, A.V., Saranin, A.A., Kasyanova, T.V., Cherevik, M.A., Pisarenko, I.V., Lifshits, V.G.: Formation of the ordered array of Al magic clusters on Si(111)7$\times$7. Phys. Rev. B **66**, 165401 (2002)

Krüger, P.: Calculations of constant-energy maps for the Tl/Si(111)-(1$\times$1) surface. Private Communication (2013a)

Krüger, P.: Calculations of the k-integrated valence-band charge distribution of Tl/Si(111)-(1$\times$1) and the charge distributions of the occupied and unoccupied surface states. Private Communication (2013b)

Krüger, P.: LDA calculations including spin-orbit coupling for Tl/Si(111)-(1$\times$1). Private Communication (2013c)

Krüger, P.: Tight binding calculations of Tl/Si(111)-(1$\times$1). Private Communication (2013d)

Lai, M.Y., Wang, Y.L.: Self-organized two-dimensional lattice of magic clusters. Phys. Rev. B **64**, 241404 (2001)

LaShell, S., McDougall, B.A., Jensen, E.: Spin splitting of an Au(111) surface state band observed with angle resolved photoelectron spectroscopy. Phys. Rev. Lett. **77**, 3419 (1996)

Lee, S.S., Song, H.J., Kim, K.D., Chung, J.W., Kong, K., Ahn, D., Yi, H., Yu, B.D., Tochihara, H.: Structural and electronic properties of thallium overlayers on the Si(111)-7$\times$7 surface. Phys. Rev. B **66**, 233312 (2002)

Li, J.L., Jia, J.F., Liang, X.J., Liu, X., Wang, J.Z., Xue, Q.K., Li, Z.Q., Tse, J.S., Zhang, Z., Zhang, S.B.: Spontaneous assembly of perfectly ordered identical-size nanocluster arrays. Phys. Rev. Lett. **88**, 066101 (2002)

Lüth, H.: Solid Surfaces, Interfaces and Thin Films. Springer, Berlin (2001)

Mak, K.F., He, K., Shan, J., Heinz, T.F.: Control of valley polarization in monolayer MoS_2 by optical helicity. Nat. Nano. **7**, 494 (2012)

Mathias, S., Ruffing, A., Deicke, F., Wiesenmayer, M., Sakar, I., Bihlmayer, G., Chulkov, E.V., Koroteev, Y.M., Echenique, P.M., Bauer, M., Aeschlimann, M.: Quantum-well-induced giant spin-orbit splitting. Phys. Rev. Lett. **104**, 066802 (2010)

Miyamoto, K., Kimura, A., Okuda, T., Shimada, K., Iwasawa, H., Hayashi, H., Namatame, H., Taniguchi, M., Donath, M.: Massless or heavy due to two-fold symmetry: surface-state electrons at W(110). Phys. Rev. B **86**, 161411(R) (2012)

Mizuno, S., Mizuno, Y.O., Tochihara, H.: Structural determination of indium-induced Si(111) reconstructed surfaces by leed analysis: $(\sqrt{3} \times \sqrt{3})R30°$ and (4×1). Phys. Rev. B **67**, 195410 (2003)

Nagano, M., Kodama, A., Shishidou, T., Oguchi, T.: A first-principles study on the Rashba effect in surface systems. J. Phys. Condens. Matter **21**, 064239 (2009)

Nagao, T., Hasegawa, S., Tsuchie, K., Ino, S., Voges, C., Klos, G., Pfnür, H., Henzler, M.: Structural phase transitions of Si(111)-$(\sqrt{3}\times\sqrt{3})R30°$-Au: phase transitions in domain-wall configurations. Phys. Rev. B **57**, 10100 (1998)

Nicholls, J.M., Reihl, B., Northrup, J.E.: Unoccupied surface states revealing the Si(111) $\sqrt{3} \times \sqrt{3}$ -Al, -Ga, and -In adatom geometries. Phys. Rev. B **35**, 4137 (1987)

Nitta, J., Akazaki, T., Takayanagi, H., Enoki, T.: Gate control of spin-orbit interaction in an inverted $In_{0.53}Ga_{0.47}As$/ $In_{0.52}Al_{0.48}As$ heterostructure. Phys. Rev. Lett. **78**, 1335 (1997)

Noda, T., Mizuno, S., Chung, J., Tochihara, H.: T_4 site adsorption of Tl atoms in a Si(111)-(1 × 1)-Tl structure, determined by low-energy electron diffraction analysis. Jpn. J. Appl. Phys. **42**, L319 (2003)

Nomura, M., Souma, S., Takayama, A., Sato, T., Takahashi, T., Eto, K., Segawa, K., Ando, Y.: Relationship between Fermi surface warping and out-of-plane spin polarization in topological insulators: A view from spin- and angle-resolved photoemission. Phys. Rev. B **89**, 045134 (2014)

Oguchi, T., Shishidou, T.: The surface Rashba effect: a $k \cdot p$ perturbation approach. J. Phys. Condens. Matter **21**, 092001 (2009)

Ohtsubo, Y., Hatta, S., Okuyama, H., Aruga, T.: A metallic surface state with uniaxial spin polarization on Tl/Ge(111)-(1×1). J. Phys. Condens. Matter **24**, 092001 (2012)

Okuda, T., Miyamoto, K., Takeichi, Y., Miyahara, H., Ogawa, M., Harasawa, A., Kimura, A., Matsuda, I., Kakizaki, A., Shishidou, T., Oguchi, T.: Large out-of-plane spin polarization in a spin-splitting one-dimensional metallic surface state on Si(557)-Au. Phys. Rev. B **82**, 161410(R) (2010)

Özkaya, S., Çakmak, M., Alkan, B.: Atomic and electronic structures of Tl/Si(111)-$\sqrt{3} \times \sqrt{3}$. Surf. Sci. **602**, 1376 (2008)

Park, C.H., Louie, S.G.: Spin polarization of photoelectrons from topological insulators. Phys. Rev. Lett. **109**, 097601 (2012)

Premper, J., Trautmann, M., Henk, J., Bruno, P.: Spin-orbit splitting in an anisotropic two-dimensional electron gas. Phys. Rev. B **76**, 073310 (2007)

Reihl, B., Schlittler, R.R., Neff, H.: Unoccupied surface state on Ag(110) as revealed by inverse photoemission. Phys. Rev. Lett. **52**, 1826 (1984)

Roushan, P., Seo, J., Parker, C.V., Hor, Y.S., Hsieh, D., Qian, D., Richardella, A., Hasan, M.Z., Cava, R.J., Yazdani, A.: Topological surface states protected from backscattering by chiral spin texture. Nature **460**, 1106 (2009)

Sakamoto, K., Eriksson, P., Ueno, N., Uhrberg, R.: Photoemission study of a thallium induced surface. Surf. Sci. **601**, 5258 (2007)

Sakamoto, K., Eriksson, P.E.J., Mizuno, S., Ueno, N., Tochihara, H., Uhrberg, R.I.G.: Core-level photoemission study of thallium adsorbed on a Si(111)-(7×7) surface: valence state of thallium and the charge state of surface Si atoms. Phys. Rev. B **74**, 075335 (2006)

Sakamoto, K., Kim, T.H., Kuzumaki, T., Müller, B., Yamamoto, Y., Minoru Ohtaka, M., Osiecki, J., Miyamoto, K., Takeichi, Y., Harasawa, A., Stolwijk, S.D., Schmidt, A.B., Fujii, J., Uhrberg, R.I.G., Donath, M., Yeom, H.W., Oda, T.: Valley spin polarization by using the extraordinary Rashba effect on silicon. Nat. Commun. **4**, 2073 (2013)

Sakamoto, K., Oda, T., Kimura, A., Miyamoto, K., Tsujikawa, M., Imai, A., Ueno, N., Namatame, H., Taniguchi, M., Eriksson, P.E.J., Uhrberg, R.I.G.: Abrupt rotation of the Rashba spin to the direction perpendicular to the surface. Phys. Rev. Lett. **102**, 096805 (2009)

Schliemann, J., Egues, J.C., Loss, D.: Nonballistic spin-field-effect transistor. Phys. Rev. Lett. **90**, 146801 (2003)

Schroder, D.K.: Surface voltage and surface photovoltage: history, theory and applications. Meas. Sci. Technol. **12**, R16 (2001)

Senftleben, O., Baumgärtner, H., Eisele, I.: Cleaning of silicon surfaces for nanotechnology. Mater. Sci. Forum **573**, 77 (2008)

Simon, E., Szilva, A., Ujfalussy, B., Lazarovits, B., Zarand, G., Szunyogh, L.: Anisotropic Rashba splitting of surface states from the admixture of bulk states: relativistic ab initio calculations and $k \cdot p$ perturbation theory. Phys. Rev. B **81**, 235438 (2010)

Sommer, C., Krüger, P., Pollmann, J.: Quasiparticle band structure of alkali-metal fluorides, oxides, and nitrides. Phys. Rev. B **85**, 165119 (2012)

Souma, S., Kosaka, K., Sato, T., Komatsu, M., Takayama, A., Takahashi, T., Kriener, M., Segawa, K., Ando, Y.: Direct measurement of the out-of-plane spin texture in the Dirac-cone surface state of a topological insulator. Phys. Rev. Lett. **106**, 216803 (2011)

Stärk, B., Krüger, P., Pollmann, J.: Magnetic anisotropy of thin Co and Ni films on diamond surfaces. Phys. Rev. B **84**, 195316 (2011)

Stolwijk, S.D., Schmidt, A.B., Donath, M., Sakamoto, K., Krüger, P.: Rotating spin and giant splitting: unoccupied surface electronic structure of Tl/Si(111). Phys. Rev. Lett. **111**, 176402 (2013)

Sugawara, K., Sato, T., Souma, S., Takahashi, T., Arai, M., Sasaki, T.: Fermi surface and anisotropic spin-orbit coupling of Sb(111) studied by angle-resolved photoemission spectroscopy. Phys. Rev. Lett. **96**, 046411 (2006)

Švec, M., Jelínek, P., Shukrynau, P., González, C., Cháb, V., Drchal, V.: Local atomic and electronic structure of the Pb/Si(111) mosaic phase: STM and ab initio study. Phys. Rev. B **77**, 125104 (2008)

Swartzentruber, B.S., Mo, Y.W., Webb, M.B., Lagally, M.G.: Scanning tunneling microscopy studies of structural disorder and steps on Si surfaces. J. Vac. Sci. Technol. A **7**, 2901 (1989)

Takaoka, K., Yoshimura, M., Yao, T., Sato, T., Sueyoshi, T., Iwatsuki, M.: Al-$\sqrt{3} \times \sqrt{3}$ domain structure on Si(111)-7×7 observed by scanning tunneling microscopy. Phys. Rev. B **48**, 5657 (1993)

Takayama, A., Sato, T., Souma, S., Takahashi, T.: Giant out-of-plane spin component and the asymmetry of spin polarization in surface Rashba states of bismuth thin film. Phys. Rev. Lett. **106**, 166401 (2011)

Takayanagi, K., Tanishiro, Y., Takahashi, S., Takahashi, M.: Structure analysis of Si(111)-7×7 reconstructed surface by transmission electron diffraction. Surf. Sci. **164**, 367 (1985)

Takeuchi, K., Suda, A., Ushioda, S.: Local variation of the work function of Cu(111) surface deduced from the low energy photoemission spectra. Surf. Sci. **489**, 100 (2001)

Tamai, A., Meevasana, W., King, P.D.C., Nicholson, C.W., de la Torre, A., Rozbicki, E., Baumberger, F.: Spin-orbit splitting of the Shockley surface state on Cu(111). Phys. Rev. B **87**, 075113 (2013)

Tung, R.T.: The physics and chemistry of the Schottky barrier height. Appl. Phys. Rev. **1**, 011304 (2014)

Vajna, S., Simon, E., Szilva, A., Palotas, K., Ujfalussy, B., Szunyogh, L.: Higher-order contributions to the Rashba-Bychkov effect with application to the Bi/Ag(111) surface alloy. Phys. Rev. B **85**, 075404 (2012)

Vitali, L., Leisenberger, F.P., Ramsey, M.G., Netzer, F.P.: Thallium overlayers on Si(111): structures of a "new" group III element. J. Vac. Sci. Technol. A **17**, 1676 (1999a)

Vitali, L., Ramsey, M., Netzer, F.: Unusual growth phenomena of group III and group V elements on Si(111) and Ge(111) surfaces. Appl. Surf. Sci. **175**, 146 (2001)

Vitali, L., Ramsey, M.G., Netzer, F.P.: Nanodot formation on the Si(111)-(7×7) surface by adatom trapping. Phys. Rev. Lett. **83**, 316 (1999b)

Wan, K., Guo, T., Ford, W., Hermanson, J.: Low-energy electron diffraction studies of Si(111)-$(\sqrt{3} \times \sqrt{3})R30°$-Bi system: observation and structural determination of two phases. Surf. Sci. **261**, 69 (1992)

Winkler, R.: Spin-Orbit Coupling Effects in Two-Dimensional Electron and Hole Systems. Springer, Berlin (2003)

Wissing, S.N.P., Eibl, C., Zumbülte, A., Schmidt, A.B., Braun, J., Minár, J., Ebert, H., Donath, M.: Rashba-type spin splitting at Au(111) beyond the Fermi level: the other part of the story. New J. Phys. **15**, 105001 (2013)

Wunderlich, J., Park, B.G., Irvine, A.C., Zârbo, L.P., Rozkotová, E., Nemec, P., Novák, V., Sinova, J., Jungwirth, T.: Spin Hall effect transistor. Science **330**, 1801 (2010)

Xiao, D., Liu, G.B., Feng, W., Xu, X., Yao, W.: Coupled spin and valley physics in monolayers of MoS_2 and other group-VI dichalcogenides. Phys. Rev. Lett. **108**, 196802 (2012)

Yaji, K., Ohtsubo, Y., Hatta, S., Okuyama, H., Miyamoto, K., Okuda, T., Kimura, A., Namatame, H., Taniguchi, M., Aruga, T.: Large Rashba spin splitting of a metallic surface-state band on a semiconductor surface. Nat. Commun. **1**, 17 (2010)

Zegenhagen, J., Patel, J.R., Freeland, P., Chen, D.M., Golovchenko, J.A., Bedrossian, P., Northrup, J.E.: X-ray standing-wave and tunneling-microscope location of gallium atoms on a silicon surface. Phys. Rev. B **39**, 1298 (1989)

Zeng, H., Dai, J., Yao, W., Xiao, D., Cui, X.: Valley polarization in MoS_2 monolayers by optical pumping. Nat. Nano. **7**, 490 (2012)

Zhang, H.M., Sakamoto, K., Hansson, G.V., Uhrberg, R.I.G.: High-temperature annealing and surface photovoltage shifts on Si(111)$7 \times 7'$. Phys. Rev. B **78**, 035318 (2008)

Žutić, I., Fabian, J., Das Sarma, S.: Spintronics: fundamentals and applications. Rev. Mod. Phys. **76**, 323 (2004)

Chapter 4
Summary

The study of Tl/Si(111)-(1×1) focused on spin-orbit-split surface states and their spin textures, which result from the intimate interplay between SOI and the surface symmetry—a current *hot topic* in condensed matter physics. The work presented in this thesis is unique in several aspects:

(i) It presents the first spin-resolved measurements of the unoccupied band structure of a heavy-metal monolayer system on a semiconducting substrate, with unprecedented agreement between theory and experiment. This has been accomplished by the development of a spin-polarized electron source, the ROSE, which enables 3D spin analysis in the inverse-photoemission experiments. The key feature of the ROSE is the rotation of the source chamber. As a result, the SR-IPE experiment allows measurements of the unoccupied electronic structure with sensitivity to two orthogonal in-plane spin-polarization directions and, for nonnormal electron incidence, to the out-of-plane spin-polarization direction, simply by rotating the source chamber. During rotation, neither the sample nor the preparation of the GaAs photoemitter are affected, so that measurements with different spin sensitivities can be conducted at the same sample preparation. Ultimately, the ROSE provides access to a wide subject area in surface science, where it is crucial to get an exact knowledge on the spin polarization of unoccupied states. Investigations on the spin-dependent unoccupied electronic structure of Tl/Si(111)-(1×1) are a case in point and demonstrate the beauty of the ROSE.

(ii) It is experimentally demonstrated that Tl/Si(111)-(1×1) features a spin-orbit-split surface state, which extends over the complete surface Brillouin zone. Tl/Si(111)-(1×1) is, thus, the prime candidate to study complex spin textures, which result from the interplay between SOI and the C_{3v} symmetry of the surface. Along $\bar{\Gamma}\bar{\mathrm{K}}$ ($\bar{\Gamma}\bar{\mathrm{K}}'$), the spin-polarization vectors of the unoccupied surface-state components rotate from the in-plane$_\perp$ spin-polarization direction around $\bar{\Gamma}$ to the direction perpendicular to the surface at the $\bar{\mathrm{K}}$ ($\bar{\mathrm{K}}'$) points. The continuation of these states are predominately out-of-plane spin polarized along $\bar{\mathrm{K}}\bar{\mathrm{M}}$ ($\bar{\mathrm{K}}'\bar{\mathrm{M}}$). Along $\bar{\Gamma}\bar{\mathrm{M}}$, the state is found to be a surface resonance, whereas symmetry arguments force the spin-polarization direction to the in-plane$_\perp$ direction. In total, experimental evidence is found, which indicate a spin chirality in reciprocal space around the $\bar{\mathrm{M}}$ point.

S.D. Stolwijk, *Spin-Orbit-Induced Spin Textures of Unoccupied Surface States on Tl/Si(111)*, Springer Theses, DOI 10.1007/978-3-319-18762-4_4

(iii) The unoccupied surface state exhibits giant spin-orbit-induced spin splittings around and at the high-symmetry points of the system. Around $\bar{M}$, splittings are observed, which show strong similarities to the predictions of the Rashba-Bychkov model. On the other hand, deviations are found with respect to the spin-polarization directions. Remarkably, the detected splittings are in the same order of magnitude as the giant splitting found for the famous Bi/Ag(111) surface alloy. At $\bar{K}$ ($\bar{K}'$), an energy splitting is found, which is related to the C_3 symmetry of the high-symmetry points. Intriguingly, the spin-dependent energy splitting amounts to 0.6 eV, more than double of the splitting of the occupied state and the largest spin-orbit splitting measured to date in any spin-orbit-dominated surface system (at and around a high-symmetry point). With the help of theoretical calculations, the giant splittings are allocated to the proximity of the unoccupied surface state to the heavy adlayer nuclei.

(iv) At $\bar{K}$ and $\bar{K}'$, almost 100 % out-of-plane but oppositely spin-polarized valleys are detected. These lie within the fundamental band gap of the Si substrate and appear close to the Fermi level. Remarkably, doping by additional Tl adsorption renders the valleys metallic, which results in a peculiar Fermi surface, where backscattering is strongly suppressed. This may be important for future spintronic or valleytronic devices.

In conclusion, the presented work demonstrates that SR-IPE with the ROSE can provide important insight into SOI-induced phenomena in systems with reduced dimensionality. Here, the strength of the experiment is, first and foremost, to resolve all necessary spin information of the unoccupied surface electronic structure. Ultimately, the presented results pave the way for ongoing studies on spin-orbit-dominated surface systems such as topological insulators and other metal surfaces or adlayer systems. One example is Tl/Ge(111)-(1×1), the isovalence electronic system to Tl/Si(111)-(1×1). Results on Tl/Ge(111)-(1×1) will be presented in the Master thesis of Eickholt (2014).[1] Like Tl/Si(111)-(1×1), Tl/Ge(111)-(1×1) features an unoccupied surface state, which can be detected throughout the complete surface Brillouin zone and exhibits complex spin textures. A highlight of Tl/Ge(111)-(1×1) is the spin texture around $\bar{M}$, which resembles but is not alike the spin chirality found on Tl/Si(111)-(1×1). In particular, a comparison of Tl/Si(111)-(1×1) with Tl/Ge(111)-(1×1) may elucidate even subtle effects, e.g., substrate-induced differences, to further unravel the consequences and the impact of SOI on surface electronic states.

Reference

Eickholt, P.: Spinstrukturen in der unbesetzten elektronischen Struktur von Tl/Ge(111). Master's thesis, Münster University (2014)

---

[1] P. Eickholt was co-supervised by the author.

Appendix

A.1 Additional SR-IPE Spectra

Here, SR-IPE spectra of Tl/Si(111)-(1×1) are shown, which are not presented within the main manuscript for reasons of clarity, but were used to determine $E(\mathbf{k}_{\parallel})$ dispersions. Figure A.1a and b shows SR-IPE spectra along $\bar{\Gamma}\bar{\text{K}}$ and $\bar{\Gamma}\bar{\text{M}}$, respectively.

A.2 $E(\mathbf{k}_{\parallel})$ False Color Plots

With the introduction of hemispherical electron analyzers in ARPES experiments, which allow a parallel detection of multiple electron emission angles, results on the occupied surface electronic structure are often presented as $E(\mathbf{k}_{\parallel})$ false color plots. In comparison with the presentation of single energy distribution curves, this has the advantage that one obtains a quick overview of the $E(\mathbf{k}_{\parallel})$ dispersion. Typically, the detection unit of the analyzer provides $E(\theta)$ images, which can be translated into $E(\mathbf{k}_{\parallel})$ plots by relating each measured point to the momentum parallel to the surface. Often, to pronounce the observed spectral features the second derivatives of the data are shown. Usually, the second derivatives are calculated numerically from smoothed raw data to produce smooth second derivatives images. It should be noted that the $E(\mathbf{k}_{\parallel})$ image of the second derivatives may convey a misleading impression of the band structure as not the real spectral intensities are shown. Unfortunately, it is not always clear whether the raw data or the second derivatives of the data are presented.

The $E(\mathbf{k}_{\parallel})$ false color plots presented in this work were obtained with a similar approach. For a set of SR-IPE spectra along one $\mathbf{k}_{\parallel}$ direction, i.e., a set of spin-resolved spectra for various angles of electron incidence, the procedure is as follows. First, the second derivatives of the SR-IPE spectra are numerically calculated for each θ and spin channel. Subsequently, a small smoothing is applied. A smoothing is chosen, which influences the data minimally and produces smooth enough second

S.D. Stolwijk, *Spin-Orbit-Induced Spin Textures of Unoccupied Surface States on Tl/Si(111)*, Springer Theses, DOI 10.1007/978-3-319-18762-4

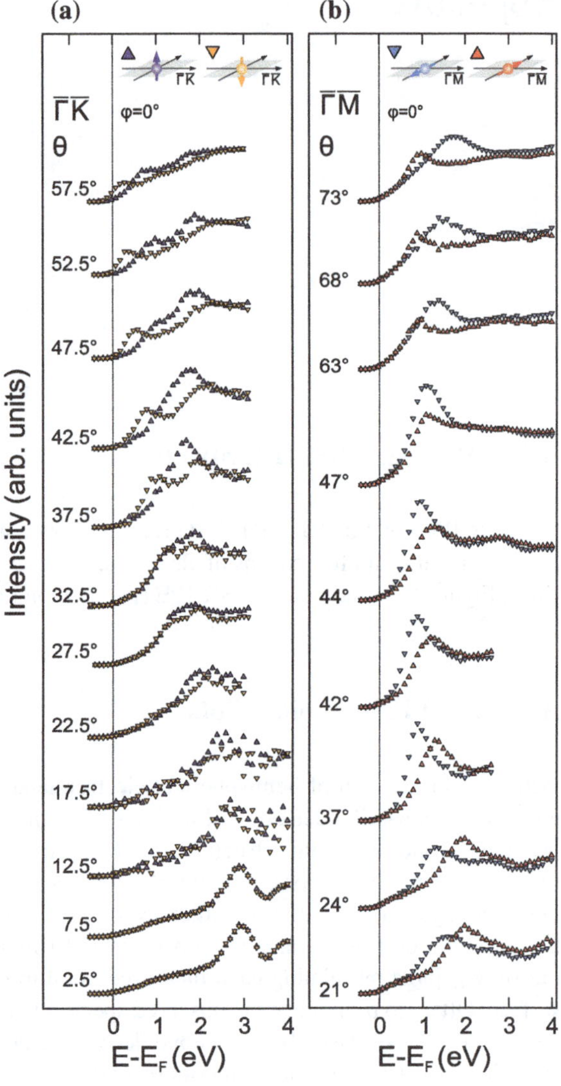

Fig. A.1 **a** SR-IPE spectra along $\bar{\Gamma}\bar{\mathrm{K}}$ with sensitivity to the out-of-plane spin-polarization direction (*purple* up-pointing and *orange* down-pointing *triangles*). **b** SR-IPE spectra along $\bar{\Gamma}\bar{\mathrm{M}}\bar{\Gamma}$ with sensitivity to the in-plane$_\perp$ spin-polarization direction (*red* up-pointing and *blue* down-pointing *triangles*). θ and φ denote the angle of electron incidence and the sample azimuth, respectively

derivatives images. Note that this procedure abstains from interpolating the data prior to the calculation of the derivatives, although these would produce even smoother images. The second derivatives are used to plot an $E(\theta)$ image. To account for the limited angular resolution $\Delta\theta$ of the experiment, the $E(\theta)$ data is convoluted with a Gaussian function (FWHM= $\Delta\theta$) along the angular axis. $E(\theta)$ is transformed into an $E(\mathbf{k}_{\|})$ false color plot by calculating $\mathbf{k}_{\|}$ for each point.

Curriculum Vitae

Personal Information

First Name/Surname:	Sebastian David Stolwijk
Date of Birth:	19.04.1985
Place of Birth:	Münster
Nationality:	Dutch
E-Mail:	sebastian.stolwijk@wwu.de

School Education

08/1991–06/1995	Johannes-Grundschule in Altenberge
08/1995–06/2001	Realschule im Kreuzviertel in Münster
08/2001–06/2004	Friedensschule in Münster
06/2004	Allgemeine Hochschulreife

University Education

10/2004–10/2009	Studies of physics at the Westfälische Wilhelms-Universität Münster
10/2009	Diploma degree with distinction; diploma thesis: "Die elektronische Struktur von Yttrium-(0001)"
11/2009–04/2014	Ph.D. studies at the Westfälische Wilhelms-Universität Münster; supervisor Prof. Markus Donath
07/2014	Thesis defense

Peer-Reviewed Publications

1. Stolwijk, S.D., Schmidt, A.B., Donath, M. (2010), "*Surface state with d_{z^2} symmetry at Y(0001): A combined direct and inverse photoemission study*", Phys. Rev. B **82**, 201412(R).

S.D. Stolwijk, *Spin-Orbit-Induced Spin Textures of Unoccupied Surface States on Tl/Si(111)*, Springer Theses, DOI 10.1007/978-3-319-18762-4

2. Amann, P., Cordin, M., Redinger, J., Stolwijk, S. D., Zumbrägel, K., Donath, M., Bertel, E., Menzel, A. (2012), "*Circumstantial evidence for hydrogen-induced surface magnetism on Pd(110)*", Phys. Rev. B **85**, 094428.
3. Sakamoto, K., Kim, T.H., Kuzumaki, T., Müller, B., Yamamoto, Y., Minoru Ohtaka, M., Osiecki, J., Miyamoto, K., Takeichi, Y., Harasawa, A., Stolwijk, S.D., Schmidt, A.B., Fujii, J., Uhrberg, R.I.G., Donath, M., Yeom, H.W., Oda, T. (2013), "*Valley spin polarization by using the extraordinary Rashba effect on silicon*", Nat. Commun. **4**, 2073.
4. Stolwijk, S.D., Schmidt, A.B., Donath, M., Sakamoto, K., Krüger, P. (2013), "*Rotating spin and giant splitting: Unoccupied surface electronic structure of Tl/Si(111)*", Phys. Rev. Lett. **111**, 176402.
5. Stolwijk, S.D., Wortelen, H., Schmidt, A.B., Donath, M. (2014), "*Rotatable spin-polarized electron source for inverse-photoemission experiments*", Rev. Sci. Instrum. **85**, 013306.
6. Stolwijk, S.D., Schmidt, A.B., Sakamoto, K., Krüger, P., Donath, M. (2014), "*Thin line of a Rashba-type spin texture: Unoccupied surface resonance of Tl/Si(111) along* $\bar{\Gamma}\bar{M}$", Phys. Rev. B **90**, 161109(R).
7. Stolwijk, S.D., Sakamoto, K., Schmidt, A.B., Krüger, P., Donath, M. (2015), "*Spin texture with a twist in momentum space for Tl/Si(111)*", Phys. Rev. B **91**, in press.

Selected Conference Presentations

1. Stolwijk, S.D., Sakamoto, K., Schmidt, A.B., Krüger, P., Donath, M. (2012a), "*Tl/Si(111)—rotation of the Rashba spin perpendicular to the surface—the unoccupied electronic structure*", 29th European Conference on Surface Science (ECOSS-29), Edinburgh (Scotland).
2. Stolwijk, S.D., Sakamoto, K., Schmidt, A.B., Krüger, P., Donath, M. (2012b), "*Tl/Si(111)—rotation of the Rashba spin perpendicular to the surface—the unoccupied electronic structure*", DPG Spring Meeting, Berlin.
3. Stolwijk, S.D., Sakamoto, K., Schmidt, A.B., Krüger, P., Donath, M. (2012c), "*Tl/Si(111)—rotation of the Rashba spin perpendicular to the surface—the unoccupied electronic structure*" (poster), 16. Hiroshima International Symposium on Synchrotron Radiation, Hiroshima (Japan).
4. Stolwijk, S.D., Sakamoto, K., Schmidt, A.B., Krüger, P., Donath, M. (2013a), "*Unoccupied surface state of Tl/Si(111): Rotating spin and giant splitting*", DPG Spring Meeting, Regensburg (Germany).
5. Stolwijk, S.D., Sakamoto, K., Schmidt, A.B., Krüger, P., Donath, M. (2014), "*Giant splitting of unoccupied surface resonant state on Tl/Si(111)*", DPG Spring Meeting, Dresden (Germany).

Award

1. Student Poster Award of the 16. Hiroshima International Symposium on Synchrotron Radiation, Hiroshima (Japan), 2012.

Zeitfracht Medien GmbH
Ferdinand-Jühlke-Straße 7
99095 Erfurt, Deutschland
produktsicherheit@kolibri360.de